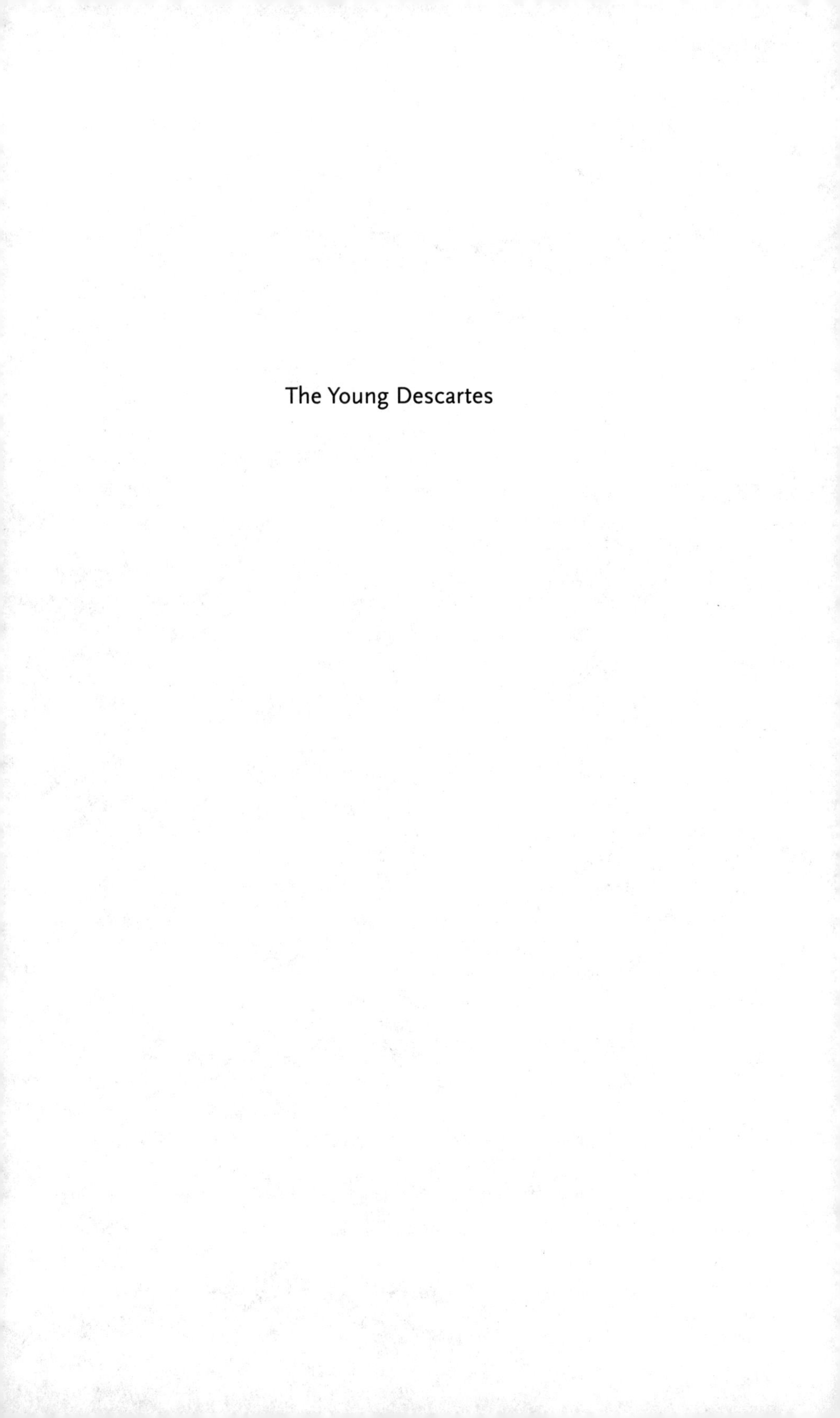

The Young Descartes

The Young Descartes

Nobility, Rumor, and War

HAROLD J. COOK

The University of Chicago Press
Chicago and London

The University of Chicago Press, Chicago 60637
The University of Chicago Press, Ltd., London

Published 2018
Printed in the United States of America

27 26 25 24 23 22 21 20 19 18   1 2 3 4 5

ISBN-13: 978-0-226-46296-7 (cloth)
ISBN-13: 978-0-226-54009-2 (e-book)
DOI: https://doi.org/10.7208/chicago/9780226540092.001.0001

Published with support of the Susan E. Abrams Fund.

Library of Congress Cataloging-in-Publication Data

Names: Cook, Harold John, author.
Title: The young Descartes : nobility, rumor, and war / Harold J. Cook.
Description: Chicago : The University of Chicago Press, 2018. | Includes bibliographical references and index.
Identifiers: LCCN 2017025412 | ISBN 9780226462967 (cloth : alk. paper) | ISBN 9780226540092 (e-book)
Subjects: LCSH: Descartes, René, 1596–1650. | Philosophers—France—Biography.
Classification: LCC B1873 .C66 2018 | DDC 194 [B]—dc23
LC record available at https://lccn.loc.gov/2017025412

♾ This paper meets the requirements of ANSI/NISO Z39.48–1992 (Permanence of Paper).

Dedicated to former President Ruth Simmons,
Provost David Kertzer,
and Dean Rajiv Vohra and their successors
Who made this journey possible

CONTENTS

PREFACE

René Descartes has long been a problem for me. His name is certainly famous, popping up in popular as well as highbrow literature as one of the chief founders of modern philosophy and science or a precursor of the Enlightenment. The usual story treats him as a solitary figure who, perhaps with the help of lying in bed well into the late morning hours, worked out the metaphysics of nature by thinking hard: a genius deducing eternal truths from first principles. Yet he himself said that human beings exist as mind and body united. If we put his mind back into this body, and his body into the midst of his world, whom would we find? He seems to have authored a lost treatise on fencing, and according to all the biographers who wrote about him in the seventeenth century, he was present in arms at the Battle of the White Mountain, often considered the opening conflict in what would turn out to be the Thirty Years' War. Was he simply a philosopher, then?

There are other intriguing problems. Descartes is understandably considered French, or a French cosmopolitan, and he loved Paris, where his closest friends could be found. Yet for his last twenty years and more, he lived away from his home country. Why? He spent almost all of that time abroad in the Dutch Republic, and that is where he wrote and published the books that made him famous. He benefited much from the information he gathered from his Dutch acquaintances as well as from their questions and encouragement, and he acquired a bit of spoken Dutch, although he apparently never became fully comfortable in it. So should we think of him as not simply a French philosopher? And how was he supporting himself? He had sold off what he inherited from his mother years before, held no public office, didn't seem to take pupils, and was without a visible patron. Even in The Netherlands, however, he kept moving from place to place. Moreover, while many observers see him as a good son of the Catholic Church, some of his early friends turn out to have been freethinking libertines, he expressed affection for people who were Calvinist, and several of his most active supporters in later years belonged to marginalized Catholic groups, such as the Oratorians and Jansenists. He finally departed from Amsterdam on another journey, to the court of the freethinking Queen Christina of Sweden, where he died in February 1650. She sent one of her largest warships to convey him to Stockholm. That extraordinary mark of distinction is very curious, suggesting that she considered him more than an ordinary visitor. Did she imagine him to be a simple philosopher, or some other kind of person?

Perhaps Descartes's early years, in the period before he became known for his printed words, can shed some light on these and other events that do not fit comfortably in the consensus view of his life. His biographers have for the most part brushed aside his first thirty-five years, when he was constantly in motion, as preliminary to the important business of writing philosophy. During the early years, the dapper young nobleman introduced himself to strangers as the sieur du Perron and served on the battlefields of central Europe before traveling on to Italy and making the acquaintance of high-ranking figures in the church. Moreover, if one is attentive to Descartes's own writings, further questions emerge that might need explanation, such as his views about human and animal physiology and the passions, which did not privilege the male. His mother's family saw to his education, and his political alliances seem to have been with women: possibly the duchesse du Chevreuse, the queen regent Marie de Medici, and Queen Anne; certainly Princess Elizabeth and Queen Christina. Apparently the female connections in his life require as much attention as the male ones.

In short, the more one considers Descartes's early life, his world, and his writings, the more questions arise. Having myself moved into a privileged environment where the pursuit of knowledge and education remain valued ideals, I decided to try to look for him in the seventeenth century. He turns out to have been right in the midst of some of the most important struggles for Europe in his time.

So, to come back to the question of why Descartes spent the second half of his life in The Netherlands: I now think that he had become an exile.

From my earlier studies on The Netherlands, in which I was alert to his presence there and was guided by many fine intellectual historians, I had come to doubt many of the messages I had previously learned about Descartes. His Dutch friends seem to have thought of him differently than the later legends. Why was he there, I began to wonder. So I began picking up parts of his story before he settled in the Dutch Republic. As I did so, some of the oddest reports appeared to become more plausible than not, suggesting further reading. Then, paging through indexes to the papers of Cardinal Richelieu—the chief minister to the French king Louis XIII—I stumbled across a reference to one "Descart" as a figure in an action at the siege of La Rochelle: could this be our man? If so, it would confirm the statements of his first biographers, which have been doubted in recent generations. Similarly, when reading an authoritative account of the reign of Louis XIII, I learned that his father had been one of the chief judges in a political trial—a kind of kangaroo court—that passed the death sentence on one of Richelieu's enemies, a young aristocrat, the comte de Chalais. That fact can be verified, and contextualized. Chalais had been invited into a grand plot by his lover, a leading member of the aristocracy, Marie de Rohan, the duchesse du Chevreuse. Her father in turn was the lord of the town in which Descartes was born as well as the person in charge of the greatest event at René's boyhood school: the ceremonial entombment of the heart of the assassinated king, Henri IV, above the chapel altar. Moreover, Descartes's most authoritative early biographer says that he had sought the patronage of the person who happened to be Marie de Rohan's second husband, who also came from one of the leading families of France and the Holy Roman Empire, the Guises of Lorraine. Their son in turn later translated Descartes's *Meditations* from Latin into French, with the author's help, and Marie and her son later became major patrons of the religiously unorthodox Jansenists, who held Augustinian views on predestination that, to my surprise, Descartes also advocated. The trial of the comte de Chalais seems to have coincided with Descartes's final break with his father, which would not be surprising if they were on different sides in this important event in the struggle for power in France.

But is it plausible that a philosopher who is usually said to be eager for privacy had connections with events at the great siege, or to grand persons related to a deadly conspiracy? Well, for his father to be sitting on a court formed by Richelieu to dispose of an enemy signals that he, at least, was far more than a simple lawyer. In fact, Descartes's father turns out to have been an important member of the administration of government in the period, a status also characterizing most of René's own friends and acquaintances. A bit of additional scratching further highlights how in his own day René Descartes would have seen himself not as bourgeois but noble. Indeed, his family name came from one of the estates acquired by his great-grandfather—the Great René—in the political warfare of the early sixteenth century, a place called Cartes. (In other words, *Descartes* means "from Cartes.") His mother's family, too, used titles restricted to the minor nobility. Yet another relative continued to be associated with the royal army, providing finance and supply. So we should not be surprised after all that if the young Descartes looked to make his way in the world, he would have looked to law, administration, war, or service to ruling families. Perhaps the military connection also explains his interest in mathematics? It turns out that military engineering was one of the most advanced technical enterprises of the period, requiring its practitioners to become comfortable with advanced mathematical methods and instruments, and sometimes with encryption, which used methods of substitution similar to modern mathematical notation. Descartes's older contemporary, Galileo, had first made a name for himself by teaching military engineering and by developing and selling a remarkable device he called his Military Compass, which could function like a slide rule to make calculations. A search of the literature on Descartes turns up no evidence that he had learned anything but common mathematics in school, but his first visit to The Netherlands was to learn the art of war, which at the time meant military engineering. It was when studying that subject that Descartes met the person who is often said to have put his feet on the road to physicomathematical studies, Isaac Beeckman. Such studies would also help explain what he was doing during some of his time in service in the imperial armies of the Thirty Years' War.

Moreover, if Descartes started his postschool years as a member of the minor nobility in search of military skills and aristocratic connections, then his friendship with his earliest known correspondent (other than Beeckman), Guez de Balzac, takes on new significance. Balzac is best known as a literary figure who in his youth was associated with some of the most prominent libertines of the period. He came from the same kind of family as Descartes and was of the same age, but he also had connections among some of the highest-ranking nobles at the court of queen Marie de Medici. It is well documented that Balzac later ended up on the wrong side of Cardinal Richelieu and had to flee into exile in the countryside: that happens to be the same moment when Descartes left Paris for what became his twenty-year period in the Dutch Republic. Was it a mere coincidence? Descartes had recently acted as his friend's champion, defending him against dangerous critics, but after the two went their separate ways, they would not write to each other for three years, out of fear. When at last they did write, their correspondence—if read in light of the deadly factionalism of aristocratic France—confirms that Descartes resisted returning to his homeland during the period of Richelieu's greatest power.

There is much more, too, when the evidence from Descartes's early biographers starts to be traced back into the period rather than dismissed as improbable for a studious intellectual. Although much remains in the shadows, one can reconstruct enough with reasonable probability to propose why the young nobleman Descartes came to be closely associated with certain aristocratic factions in France before fleeing Cardinal Richelieu. Only after the cardinal's death, under the regency of Queen Anne, would he return for short visits to the country of his birth. Conflicting interests in France and abroad nevertheless continued to bind him by law, war, and diplomacy. They engaged him in dangerous confrontations about God and nature, too, for concepts and beliefs were as much at stake as personal loyalties. He did not start his adult years intending to write philosophy, then: indeed, he explicitly rejected the kind of learning he had acquired in school, only much later figuring out how to speak up in print. He could not raise an army, but he could recruit the powers of nature and of the heart in an effort to put things aright. Whatever else people take away from this account, I trust that his life on the move shows Descartes to have been no armchair philosopher.

I confess, too, to having a larger agenda. Sometimes historical progress is seen as inevitable, with Descartes himself often depicted as intending to help the modern world come into being simply by writing what he thought, all being very straightforward. But in fact, people are shaped by the constraints as well as the opportunities in their lives, with few things feeling inevitable when one is living through them. I have long been interested in exploring how those real felt lives affected the ideas and practices of various "thinkers." In Descartes's case, he was, I think, vulnerable. He was socially high-ranking enough to be on the edges of power, but he was never secure; in seeking his fortune, he seriously endangered his life on more than one occasion, and he walked through rooms alert to the personal passions that lay behind the rise and fall of favorites. Why would someone living in such a world help create the new science? Many of the most exciting arguments among historians of science in the late twentieth and early twenty-first centuries have explored the importance of trust among people who knew one another well, considering those personal relationships to be built into the methods used to produce knowledge. But historians have long noticed, too, that the period in which Descartes lived was replete with uncertainty and personal distrust: deadly famine and plague ran through whole regions at a time, war was rampant and made even more horrifying by the brutalities evoked by religious and ideological division, the hunting of witches and heretics terrorized many lands, fortunes might be lost in an instant if ships went down (as they did) or if princes failed to pay what was owed (as they did), honor and office could evaporate overnight on whispers in ears. That was Descartes's world.

And yet his attempt to advance a new science emerged from the same place. Might he have been trying to resolve the reasons for doubt and mistrust? Might many of his thoughts have been provoked by the materialistic libertinism of his youth? Might his solutions have something to do with methods he became familiar with on his travels, such as engineering, medicine, and other practical sciences? The practices of those arts had to stand up to extreme conditions. Weren't the ideologues in fact the false dreamers of his world, inevitably losing their way in pursuit of their "isms," while he stripped all that away to focus on the real? His confidence in physicomathematics to understand the "how" questions, with the "why" questions left aside, might have something to do with the felt need to find a reliable foundation for assurance about truths of nature, at least, in a world of enormous stress. In other words, like so many others, the hardheaded but not hard-hearted young aristocrat came to want truth to be based on proof rather than opinion or rank. His realism also included recognition of the place of the passions in setting the universe in motion. His life and work do not therefore hover above his world but emerge from it. Such a view helps me, at least, to better understand the aims expressed both directly and indirectly in his own written works.

To make that case, however, we must venture into the kind of history that often depends on indirect evidence. I draw much of it from seventeenth-century accounts of Descartes's life because their authors had access to evidence no longer extant, but many of their declarations and hints can be said to be likely or unlikely in light of other sources. Since I am not myself trained as a historian of philosophy, much less as a Cartesian, nor as a historian of France (despite some coursework years ago), it may be that more expert scholars think that my paths of inquiry end in blind alleys. But I have tried to find the traces of interconnecting passages known to have existed in his day in order to make sense of one young man's movements, relying with gratitude on those who have scouted the ways before. In the process I have not only learned much that is new to me but stumbled across a few fresh tracks. The foray is meant to raise questions as much as to provide answers, inviting others to attend anew to a moment whose echoes can still be heard.

Beware, though: for the routes he took lead onto ground that has been hotly contested. Old castle walls may now be abandoned, the gates only needing a push to fall open, but if we dare walk through them, we need to prepare ourselves for the confusion of struggles in earnest for possession of the keep. It remains occupied. In the meantime, I have tried to follow Descartes's own advice: to do one's best and have no regrets.

PART 1

Mysteries

Remains of a Hidden Life

The name *Descartes* is now associated with intellectual summits. For many people, the name evokes the fierce engagement of French philosophy or the rise of modern science; for others, simply the famous phrase *cogito ergo sum*, "I think, therefore I am." That pithy phrase both challenges credulity and asserts the real, asking us to reexamine the foundations of everything we know on the simple basis of a proof of our existence. Over the generations, then, the process of interpreting and memorializing Descartes's works has built up a legacy of mountainous proportions. For almost anyone engaged in exploring intellectual terrains, whatever the viewpoint sought, the peak named after him is seldom out of sight. Great climbers have scaled it from every possible approach, challenging one another with feats of close reading and command of the literature; for the rest of us, all kinds of guides are ready to help any amateur to the top and back by well-tended paths.

But mountaintops best serve as a place from which to look out into the distance. If we want to get a good look at the heights themselves, we must circle back lower down and look up. If we do, we notice around us that not all the trails lead toward the top. On these lower slopes the paths move through woods and meadows, more often heading around the mountain than upward to the rock and ice. In the foothills, too, it is possible to move about more freely, without need of a fixed itinerary and someone else's rented equipment. If we pause to explore these foothills, we glimpse traces of an older geology. An earlier version of this peak rested on hills and valleys that can still be identified, and in them can be found the remains of once mighty cities. It was a wilder world, where beyond the walls wolves still roamed and armies gathered, one that frequently stirred the blood. If you have the time to nose about down here, you might even begin to wonder about the mortal and life-size person after whom the mountain came to be named, and at which of the ancient inns he might have stopped for rest and conversation with friends. It would be worth looking in, for it is said that he could be a charming gentleman, on most occasions putting his sword to the side.

In fact, the living René Descartes could have walked straight out of the pages of Alexandre Dumas's *The Three Musketeers*. Descartes took part in the action at the siege of La Rochelle, which figures prominently in the last pages of the novel, and when René introduced himself to others, he gave his name as the sieur du Perron—a title originating from an estate he inherited from his mother—while his printed portrait bore his coat of arms (see fig. 1).[1] While he received an education suitable for the French elite, he also learned to fence well and to ride skillfully, being trained to ride the "great horse" used in battle, the destrier. By his twentieth birthday he was living in Paris, where he dressed fashionably in green taffeta with a plume in his hat and a rapier at his belt, cultivating crowds of friends; enjoying love stories, music, poetry, ancient mythology, and jokes; and for a period spending long hours at the gaming tables, apparently accumulating large winnings in the process. (In French, the words *des Cartes*, which is how Descartes wrote his surname, might even suggest playing cards.) Like Dumas's protagonist, D'Artagnan, he did not walk through the city followed by mobs of retainers—as did the great aristocrats of the day—but singly, employing only a valet and a few lackeys.[2] But soon, and lasting for a decade, he would be caught up in the wars of his day, finding himself not always on the side of the victors.

Figure 1. René Descartes as chevalier, as represented in his first major biography. Frontispiece from Adrien Baillet, *La Vie de Monsieur Des-Cartes* (1691), engraved by Gerard Edelinck, after Frans Hals. © RMN-Grand Palais / Art Resource, New York.

Unsurprisingly, then, the real Descartes learned to use his sword skillfully. An early biographer who had known Descartes in person, Nicholas-Joseph Poisson, insisted on Descartes's long experience as a young soldier, "serving Mars before Minerva," and reporting that those who had known him had heard tales of his adventures from his own mouth; in Poisson's day, a personal memoir from his period of military service still existed.[3] If his seventeenth-century biographers are correct, the young soldier also somehow managed to survive the later slaughter of the imperial army at Nové Zámky (in what is now southwestern Slovakia; at the time also known as Neuhausel). Another early biographer, Pierre Borel, therefore made a point of writing that Descartes was "good both at the Pen and the Pike," and that he loved "the valiant as well as the prudent and learned."[4] That he later became known for his books places him among such other famous soldier-authors of the period as Cyrano de Bergerac. Like Cyrano, so many men in arms took an interest in the new sciences and medicine of the day that one historian has used the term "soldier-savant."[5] We should not be surprised, then, to find Descartes writing to a friend that the pursuit of true philosophy was a work of valor, or that the struggles by which one arrives at truth are like the battles in war.[6]

Put another way, like other young cavaliers of his era, Descartes cultivated chivalry. He well knew the genre's greatest model, *Amadis of Gaul*, the book so admired by Miguel de Cervantes that his own *Don Quixote* mocked the imitators. So popular was *Amadis* in the France of Descartes's youth that one of the most frequently reprinted handbooks was composed of extracts of courtly speeches from it, suitable for any occasion.[7] In the story, Amadis was secretly fathered by the union of the beautiful Elisena and the powerful king Perion, then like Moses set on the waters to be discovered and brought up in another household. There he showed his innate virtue in ignorance of his heritage, becoming a knight errant, suffering many trials and enjoying many rewards in love before eventually being recognized as King Perion's true son and rescuing the kingdom. The sieur du Perron, too, was a virtual orphan, raised by others, spending many years traveling abroad to seek his fortune and suffering pleasures and pains in the process. As far as we know, he never saved a kingdom, but he kept the faith. Poisson tells of a moment when Descartes was returning to Paris on the Orléans road and was ambushed by a rival: not only did he defend himself well, he even disarmed his opponent; then, instead of giving the killing blow, he returned the captured sword for the sake of the eyes of the lady involved.[8]

Like D'Artagnan, then, Descartes not only studied war but "delighted to discourse with Women," as Borel put it.[9] A recently discovered portrait of him as a young man certainly shows him to have been handsome, and he seems to have exhibited a great deal of personal charm when he wished. He later fathered a child out of wedlock, and he loved this daughter, Francine, very much, being greatly moved by her death at the age of five. He must also have had some feeling for his daughter's mother—Helena Jans, a printer's housemaid—since when she finally married someone else, he put up a substantial dowry for her.[10] But Descartes was best known for his conversations with women of noble rank. He had a long relationship with the young Princess Elizabeth of the Palatinate—niece of the king of England—and a shorter one with the even younger Queen Christina of Sweden. The last work he published during his lifetime, *The Passions* (1649), owed an enormous debt to the comments and inquiries of them both, especially Elizabeth. He listened. It is written with a view to the universal passions that move us all, including love and generosity. Descartes's first book, the *Discours* (1637), was originally written in French so that it could be appreciated by women,[11] and his work came to have many supporters among the ladies who frequented the salons of Paris, who helped establish Descartes's reputation as a brilliant philosopher.[12] Even better, the figure who pulls the strings in Dumas's novel (without ever making a personal appearance) is the Duchesse du Chevreuse (see fig. 2), a most remarkable woman, involved in many of the greatest events and conspiracies of the period, considered not only a charismatic beauty of many great love affairs but clearly possessed of a forceful mind, a powerful spirit, and a

Figure 2. Marie de Rohan, duchesse du Chevreuse, as the goddess Diana, having captured the stag, Charles, duc du Lorraine. Painting by Claude Deruet, Châteaux de Versailles et de Trianon. © RMN-Grand Palais / Art Resource, New York.

clear sense of the right and just; it was her son, Louis-Charles d'Albert, the duc de Luynes, who translated Descartes's *Meditations* (1641) from Latin into French, with Descartes's help.[13] Our chevalier probably knew the brilliant duchess in person.

But he was also good with men. His letters are full of affection for his male friends, even passionate expressions of love.[14] Some of his best friends were accused of moral *libertinism*, sometimes a code word for male–male love. During his exile in The Netherlands, Descartes lived for long periods with men, most important among them Étienne de Villebressieu and the abbé Claude Picot, the latter described as Descartes's agent "concerning his domestic affairs and revenues."[15] Picot, in turn, was a good friend of the poet Jacques Vallée, sieur des Barreaux, known as an epicurean and lover of the scandalous poet Théophile de Viau. About the same age as Descartes, Barreaux had been a classmate of René's at school. Barreaux's circle of friends later included not only Descartes's friend Guez de Balzac but also Claude Emmanuel Lhuillier, known as Chapelle—one of Cyrano's intimates—who was born out of wedlock to Marie Chanut, sister of the French ambassador to Sweden in whose house Descartes died. A member of Picot's family was Paul Scarron, who also became an abbé, at age nineteen, and was reputed to be a "procureur" for Louis XIII and a client of the king's chief mistress, Marie de Hautefort (one of the few women in whom Louis took an interest).[16] When Descartes paid a brief visit to Paris in 1647, he resided in the same urban palace (*hôtel*) as the family of Madame Scarron de Nandiné—whose first husband was that same Paul Scarron—better known in later life as Madame de Maintenon, mistress and then wife of King Louis XIV.[17]

Descartes is discreet about it all. But in 1643, one of his Dutch adversaries would publish a book meant to harm him, among other things attacking Descartes for his sexual "dissolution and debauchery."[18] That some of his closest friends, at least, engaged in both what would now be called homosexual and heterosexual relationships should come as no surprise, since such mixtures were common among the rulers of the kingdom as well as elsewhere. A generation earlier, the aggressively effeminate king Henri III had provoked a succession crisis when his assassination left no heir. In Descartes's own lifetime, Louis XIII (see fig. 3) was known to have favored many young men among his "pretties" (*mignons*), while the king's younger brother, Gaston d'Orléans, and a prince of the blood, the Grande Condé, were among the many male aristocrats who had intimate relations with men as well as women, often loving re-

Figure 3. King Louis XIII, armored, decorated, and with the chivalric medal of the Order of the Holy Spirit. Painted by Philippe de Champaigne, Château, Fontainebleau.

lationships among men of valor and martial skill.[19] Descartes's own portrait was painted by Simon Vouet, who depicted many of the young men around Louis XIII. Probably for such people Descartes later wrote of the love and affection of an "honorable man for his friend or mistress."[20]

Moreover, in the literature of the period not only male love but also androgyny is well represented. Perhaps the most successful work in the period of Descartes's youth was a book composed by Honoré d'Urfé, Marquis de Valromey and Comte de Châteauneuf, who died on campaign in 1625: it was titled *L'Astrée*, after the goddess of peace, and published in five installments between 1607 and 1628. While the name of one of its many characters, Celadon, came to be a byword for love, that love embodied a symmetrical sexuality based on the pre-Edenic and alchemical myth of a united person who had no distinct gender, the desired union re-creating the Androgyn.[21] The famous author of fables, Jean de La Fontaine, later called the free-swinging *L'Astrée* the "livre d'amour par excellence."[22] Gaston d'Orléans himself was utterly captivated by these tales of love, as were many of Descartes's friends.

Had Descartes's life taken a slightly different course, then, we might have seen him rise to be among the courtiers of the age, perhaps a kind of lesser duc de La Rochefoucauld, whose keen-sighted maxims exposed the springs of human action and self-regard for all to see. But Descartes ultimately chose to explain the physical sources of our embodied lives. And there is another difference: instead of ending his life amid the aristocratic households of France, Descartes died elsewhere, after two decades away from his beloved Paris. Earlier in life, his potential patrons fell from power, causing him to remain in search of opportunities; and just when his name began to be widely known, he found himself on the wrong side of the king's coldhearted chief minister, Cardinal Richelieu, and fled. His honors therefore came posthumously, and not for his actions or counsel but for his writings. Since he lived abroad, his compatriots came to know him mainly through his books and correspondence, thinking of him as a person apart, distant, alone.

Descartes has therefore earned a reputation not as an aristocrat but as a hermit. Depictions of him have consequently been shaped not by the genre of chivalry but of martyrology. They report that Descartes died tragically, an accidental victim of a woman's vanity. He is said to have risen in the wee hours of the night in the frigid dark of a Swedish winter to tutor the young queen, herself an early riser, thereby exhausting himself and dying of a respiratory illness. But it seems that in fact the queen met him in person only four or five times, while at the time of his death a serious fever was running through the court. After his death, rumors circulated that he had been poisoned,[23] while another story had it that he faked his death to run off and live among the Lapps, who were then reputed to be the most powerful magicians in the world.[24]

If not romance or martyrology, then, perhaps mystery. Mystery accumulated around him even in his lifetime. As one of his most orthodox modern biographers admits, "even though he left confusion in his wake, René Descartes always succeeded in covering his tracks."[25] Why he traveled abroad so often certainly perplexed his friends. And then there is the secrecy. When as a young man he headed off into central Europe, he jotted down in his notebook "I go masked" (*Larvatus prodeo*). It was a common phrase among people of his rank, who lived daily at court by acting out their public personae.[26] But he spent years on the road throughout central and northern Europe followed by two years in Italy, and even after he returned a second time to the Dutch Republic, he could not settle, shifting his residence from such cities as Dordrecht to Amsterdam, to Franeker, Leiden, Deventer, and Utrecht, and then even to smaller towns including Sandpoort, Endgeest, and Egmond. He dissuaded friends from Paris from coming to visit and urged his correspondent Marin Mersenne not to let people know his address, while he also used drop boxes, instructing Mersenne to contact him by sending messages to other people from whom he could collect his post.[27] He later adopted a motto from Ovid as his own: "He who lives hidden, lives well" (*Bene qui latuit, bene vixit*).[28]

The same motto is also said to have been adopted by the secretive brethren of the Rosy Cross, or Rosicrucians.[29] Coincidentally, when signing letters, he always wrote his name as "René des Cartes," abbreviated as "R.C." The fifth of the six rules of the brotherhood instructs its members to use "C.R." or "R.C." as their mark.[30] Coincidentally, too, the first of its rules commands them to profess no other thing than curing the sick without a fee, while others ask them to blend in with the people of whatever country they inhabit; our R.C. was interested in medicine from an early date, while one of his moral maxims was to live according to the customs of the region in which he found himself. There is good evidence to show that he spent time at Ulm with the noted German mathematician and reputed Rosicrucian Johannes Faulhaber, and Descartes's early notebook contains a sketch of a plan to publish a work dedicated to the Rosicrucians under the pen name Polybius Cosmopolitanus.[31] After returning to Paris from the wars in Europe, he arrived in the midst of a public uproar about subversion by the Rosicrucian brotherhood, with many people fearing that he was one of its underground leaders. Reputed Rosicrucians were numbered among some of his closest Dutch friends as well. It is this line of association that a Jesuit critic later drew on in a satire that has Descartes making visits in spirit to the moon, stars, and even outer reaches of the universe after smoking strong tobacco mixed with a secret herb.[32] Here we have an occult Descartes, far more resembling the secretive and reputedly long-lived Count Saint Germain than a straitlaced soldier or philosopher.

Who was he, then? Can we can spot him here in the foothills before attempting to climb higher?

Words on Paper

We need not doubt that about four hundred years ago the living René Descartes authored several books that became enormously influential in his own day and long after: in fact, Descartes is usually thought to be identical to his authorial persona. The power of his philosophy stems in part from his style of writing, which is simple and direct, not disputing with other authors or arcane opinions but setting out the real world as plainly as he could. In the first book he published, the *Discours*, he goes so far as to address the reader personally, almost as if writing a confessional letter to a distant friend. He begins by taking us aside to share a gentle joke about the human race, observing that good sense seems to be "the best distributed thing in the world," since everyone thinks they have so much of it that they don't need any more.[33]

But perhaps the joke is on us. Since "they" are the human race, we are them, and perhaps we are not so mistaken about having good sense after all. He himself has no special abilities, he says, since he shares the disposition of the rest of us: "For my part, I have never presumed my mind to be in any way more perfect than that of the ordinary person."[34] If we all have ordinarily reasonable minds, then, "the diversity of our opinions" must be due not to some people being intelligent and others not but rather to our attending to different things. We are diverse because we have different life experiences and different interests. If we want to find agreement, then, we need to proceed with deliberation, drawing demonstrative conclusions at each stage only with care. For himself, he had the good fortune to happen upon some paths "in my youth" that led him to search out the truth in a way he thought was solid and important enough to share. And yet, he acknowledged, "I may be wrong: perhaps what I take for gold and diamonds is nothing but a bit of copper and glass." He is not setting out directions for everyone, then, simply writing about how it came to be that he tried "to direct my own" mind, sharing a kind of personal history, or even a fable, which some people might find useful and no one should find harmful. He hoped only that his readers would thank him for setting down his views honestly.[35]

As a reader, then, you meet the author as a mature, accomplished but humble person, a sharp observer of humanity capable of making jokes at his own expense, who invites us to partake in a quiet conversation about how things really are. You are free to stay or leave as you like. Descartes's persona comes across not only as thoughtful and cheerful but as personal, reassuring, and inviting. He goes on to offer a few incidental details about himself. But he is also discreet. His chief works only occasionally address opponents, and then only with generalized passing swipes, seldom dropping names or referring to details, revealing only what is common to all and what is required to persuade the reader to listen in. We are included in a conversation removed from the hubbub of business and domestic life, or the fawning and intimidation of serving those in power, or the classroom. We are sitting apart, with time enough, the words alone having importance. Descartes was well versed in the methods of his literary friends, who cultivated the ability to invite their readers into dialogues of the imagination.

But there are other kinds of writing left behind by the living Descartes. As one would expect of evidence from four centuries ago, many sources have gone missing. In his case, however, that was not due to neglect alone but also to deliberate disposal. Descartes himself seems to have secreted away many of his personal papers, threatening his acquaintance Mersenne that "if I do not die in my own good time and in a good humor with the men who remain living, they will certainly not see [my papers] for more than a hundred years after my death."[36] In fact, he did die unexpectedly, and not among friends. What he might have buried deep will never be known. But what he openly left behind was sifted and culled by other people, too. Consequently, some papers were lost while others appear to have been destroyed. For instance, he seems to have written about his experience in arms, but it is now missing. His *Discours* promised a further work on the nature of the soul, which was never published nor recorded among his surviving papers. All this leaves plenty of room for speculation about what else disappeared.

Hints about the fate of his papers come chiefly from Descartes's first scholarly biographer, Adrien Baillet,[37] supplemented by recent investigations by the editors of a new edition of the Descartes correspondence.[38] We know of two main caches of papers surviving his death. When he prepared for his departure from The Netherlands to Sweden in 1649, he put the papers and other items he wished to take along in a couple of chests for shipment, while another batch of his papers was deposited in a case left behind in Leiden with his trusted friend, the physician Cornelis van Hogelande (a fellow Catholic and reputed Rosicrucian).[39] Descartes also left instructions with Hogelande on how to proceed should he die abroad. In such a case, Hogelande and a discreet friend or two could go through the papers, since there was nothing secret in them (*qu'il y ait rien de secret*), but because some of them might contain information that some of his correspondents would not want to become public, he would probably think it best to burn them. It is suggestive that Descartes draws a distinction between secrets and privacy, implying that he was privy to *secrets*—a word often associated with governments—and that they would not be found among the papers he was leaving behind. In any case, he continued, he only wanted posterity to have the letters that his enemy Voetius wrote to Mersenne, which Descartes thought would help protect his posthumous reputation (*& que je desire être gardées pour servir de préservatif contre ses calomnies*); these he set aside separately in the lid of the case.[40]

Once news arrived of his friend's death, Hogelande opened the case in the presence of a notary and three witnesses: a mysterious French-speaking officer of the Dutch army named Louis de la Voyette who later served the Swedes (an intelligence officer?), and two professors from the University of Leiden, the mathematician Frans van Schooten, who had collaborated with Descartes on the publication of some of his works, and the physician Johannes de Raey, another friend. De Raey was still alive at about the age of sixty-eight when Baillet was trying to find out more about his subject, and he insisted (in a message conveyed via the theologian Philipp van Limborch, one of John Locke's friends) that the letters Descartes had left in Leiden were few and unimportant. But it is also known that De Raey did not like Baillet's prying into his friend's personal life, telling Van Limborch that it had been very simple, whereas the French were falsifying it (*Vita Cartseii res est simplicissima, et Galli eam corrumperent*): perhaps De Raey wanted to throw Baillet off the scent.[41] In addition, another long-standing friend of Descartes was present, the Dutch nobleman Anthony Studler van Zurck, who had served as an early recipient of letters intended for Descartes and loaned him a large sum of money (perhaps to pay fees to the French government for a pension that never took effect).[42] Van Zurck seems to have kept at least some of the papers not destroyed, but all trace of these papers disappeared in the early nineteenth century aside from a few apparently returned by Hogeland to their originators, such as those of the secretary to the Prince of Orange, Constantijn Huygens.[43] Only a few of the letters that Descartes explicitly wanted preserved are now known, and these only from copies. Baillet suspected that most of the letters were burned.[44]

In Stockholm, following Descartes's death on February 11, 1650, another meeting was convened, this time by the French ambassador, Pierre Chanut. It included a representative of the queen, sieur Erric Sparre Baron de Croneberg. The other members could also be trusted to be discreet: Descartes's valet, Henry Schlüter, and the chaplain and the secretary of the French embassy. They went through all Descartes's possessions and made an inventory of his books and family papers (to share with his relatives). The following day—were they first handled by the queen or the baron?—the ambassador took possession of them. Chanut seems to have intended to publish at least some of the manuscripts, and probably about four years later, perhaps working with Christiaan Huygens (the famous scientist, and a Dutch Francophile), the ambassador composed a list of those he thought important for understanding Descartes's scientific ideas, since known as the "Stockholm Inventory."[45] But other papers, such as the letters of the Princess Elizabeth, which Descartes had kept separate from the others and were returned to her at her request, were not recorded, so in fact we do not know the full extent of what survived from Stockholm, either.[46]

When the Descartes family expressed no particular interest in the papers, Chanut turned a large batch of them over to his brother-in-law, another friend of Descartes, Claude Clerselier. But a boat on which they were being transported sank in the river Seine. It took three days to recover the chest in which they were contained, and then many days further to dry them out, all of which put them into great disorder.[47] Clerselier had worked with Descartes since 1644 and later devoted much effort to publishing editions of Descartes's work and correspondence (for instance, bringing out a French edition of Descartes's *Description of the Human Body* in 1664), but he has been shown to have heavily edited the letters he published, adding some passages and deleting others. Clerselier shared some of Descartes's unpublished notebooks with other scholars such as Gottfried Wilhelm Leibniz and Walter Ehrenfried Tschirnhaus, and their notes in turn provide invaluable evidence to us, but the materials on which they were based are now gone.[48]

Other hints about Descartes's life began to appear, however, many of them based on personal recollections and personal papers acquired through unknown channels. Daniel Lipstorp of Lübeck compiled recollections of Descartes he acquired from Van Schooten and De Raey along with an account from a young "disciple" of De Raey's who had met Descartes, named Van Berkel.[49] On the French side, Pierre Borel produced an important short account of Descartes's life in 1653 that went through several subsequent editions, including a translation into English in 1670.[50] Although often overlooked, it is worth attention. Borel had taught engineering at Castres, proposing a canal that would later be constructed by others as the Canal du Midi, and would go on to be known as a committed Cartesian, chemist, royal physician, and member of the Académie Royale des Sciences.[51] He added some crucial information to Lipstorp's account, not only from an epitaph written by Marcus Zuerius Boxhornius at Leiden but, more important, from the recollections of Descartes's military and engineering friends, particularly Descartes's former roommate Étienne de Villebressieu (also a physician, alchemist, and engineer). Clerselier himself published some of Descartes's letters posthumously in 1657, 1659, and 1666, and in doing so, he conveyed a few things about the life of the author in the prefaces. Finally, Queen Christina commissioned Nicholas-Joseph Poisson, an Oratorian priest, to honor her philosopher, and he published two studies of Descartes's ideas that contain snippets of biographical information. Although he never completed a full biography, Poisson seems to have handed over some materials (originally obtained from the queen?) to Clerselier.[52]

But the chief source of information from the period, based largely on documents no longer extant, comes from the biographical project associated with the name of Adrien Baillet. When Clerselier died in 1684, having collected masses of Descartes materials but without having produced the biography he had hoped to write, the papers in his possession went to the abbé Jean-Baptiste Legrand. Legrand wrote further to people throughout France and Europe who might have additional information about Descartes, but to sort out the mass of collected materials he finally engaged a learned librarian, another abbé, Adrien Baillet, as a kind of ghost writer.[53] Baillet was then working on revisions to a huge critical encyclopedia covering all of literature and on a comprehensive study of the lives of the saints, but he was willing to be interrupted, and he was able to turn the two-volume biography over to the printer in February of 1691.[54] Two years later he published an abridgement of it that contained some further information (and which was translated into English in 1693) but he never completed the planned revised edition of the whole.[55] Legrand kept possession of the papers used by Baillet, but following Legrand's own death in 1704, they went to his mother and then disappeared.[56] We do not even have a partial inventory of what was lost.

Because he had so much material at his disposal that has since gone missing, then, Baillet's version of Descartes's life must be our primary guide. But of course it requires interpretation. The fact that the Clerselier project had been delayed for thirty years suggests that more than a lack of energy was at work: probably there were internal conflicts about how to present Descartes's life to the public. For example, one of the problems—if the hypothesis that I will outline in a moment is correct—would have been Descartes's affiliations with discontented nobles around Gaston d'Orléans. Gaston would have become king had not his nephew been born—the child who became Louis XIV and ruled when the Baillet biography was published—and many of the discontented would also take up arms against Louis's mother, the queen regent, Anne of Austria, in the midcentury rebellions known as the Fronde: at one point, she and her son had to be smuggled out of Paris at night. With the growing power of the Sun King, Descartes's apparent associations with the troubled history of the period would have made finding an acceptable version of his life quite difficult had he had any connections with Gaston or those nearby.

There were certainly problems in reconciling Descartes's philosophy with the increasingly conservative religious orthodoxy of the period, too. While Descartes wrote of God, he came very close to identifying God and Nature, a classically "atheistic" opinion. Descartes's works had also been condemned by the archbishop of Paris in 1671 for raising questions about the Eucharist, while in 1675 a declaration of the royal Grand Council banned the teaching of his works. Moreover, although Baillet was an experienced and critical editor, he was living in a moment immediately after Louis XIV's infamous revocation of the Edict of Nantes in 1685. The revocation withdrew the civil rights of Protestants in France, forcing them to flee, convert, or face confiscation of all their goods under threat of torture or execution. Descartes held the views of Augustine in high esteem, and so did Clerselier, but some of Descartes's chief advocates were members of the controversial Augustinian "sect" of Jansenists, whose teachings on grace and salvation were considered too close to Protestantism to be acceptable to most of the ruling authorities in the French church. As a member of the Catholic hierarchy working in the libraries of noblemen, Baillet would have to be very cautious about what he reported, and how he spun it, if he wanted to make Descartes conformable for his audience.

Another possibility would have made an orthodox biography of Descartes even more difficult: One of Descartes's recent biographers, Geneviève Rodis-Lewis, writes that Chanut was "known to be a member" of the Compagnie, or Société du Saint Sacrement, and Christina's biographer, Susanna Åkerman, adds that Clerselier was also a member.[57] The Compagnie was a secret society founded by the duc de Ventadour in 1627 as a way to unify France around a Catholic ethical culture. Mostly it included curés, bishops, jurists, administrators, and noblemen, including Condé's brother, the Prince of Conti. It was sometimes frequented by ethical libertines, as well, including Pierre Gassendi.[58] They advocated the prohibition of dueling, reverent behavior in churches, and other reforms, including the withdrawal of civil rights from Huguenots—Ventadour himself had attacked the Huguenots in the south of France—but they did so in private, advancing their interests not as a group but through individual members of influence who put forward the common proposals as their own. Cardinal Bérulle of the Oratorians, a group with whom Descartes himself was closely associated, had encouraged the Compagnie's formation, although the archbishop of Paris refused his own approval and Pope Urban VIII refrained from granting it any special status.[59] In due course the *parlement* (parliament) of Paris decided to prohibit secret assemblies "and the Compagnie du Saint-Sacrement gradually yielded to the pressure of government hostility and suspended operations."[60] An examination of early twentieth-century publications of archival sources of the Compagnie for any references to Chanut, Clerselier, or Descartes has drawn a blank.[61] That lack of confirmation cannot completely eliminate the possibility of a connection, but if it existed, it would have made the presentation of Descartes and his views during Clerselier's lifetime full of inner contradictions because some other contemporaries thought that his subject was privately materialist, and so atheistic.

The moment of publication also followed the effort by Louis XIV and the imperious minister Louvois to destroy the Palatinate. In 1688, on the pretext that the king's sister-in-law should inherit the principality, French armies marched into lands on the east bank of the Rhine, burning cities and towns to the ground and inaugurating what would become known as the Nine Years' War. With the French in Germany, the Dutch stadtholder William III, Prince of Orange, gained a brief moment of safety for launching his coup in England—better known as the Glorious Revolution—which then allowed him to bring Britain and the United Provinces together into the war against France. But that meant further difficulties for Baillet, since Descartes's early military service in the forces of the Prince of Orange and among a variety of German princes who now fought France would also have to be reported carefully if he were not to seem disloyal to contemporary readers.

Baillet was therefore understandably eager to highlight aspects of Descartes's life and ideas that made him acceptable in this moment of conservative reaction in politics and religion, and to play down any other associations. He only hinted at the personal connections of Descartes in the early seventeenth century that might seem troubling from the viewpoint of the religious absolutism of Louis XIV. He was successful enough in his effort to receive approval for the dedication to be made to Louis Boucherat, chancellor of France and signer of the Edict of Nantes.[62]

Baillet did not simply bow and scrape, however. He had first been widely noticed for his nine-volume *Jugements des Savans* (1685–86), which celebrated the "liberty to judge." His subsequent multivolume *Vie des Saints* (1695–1701) would show sharp criticism of the sources, questioning the documentation of reputed miracles in recent centuries.[63] His critics often accused him of being more a copyist than a historian, but his doubt about the existence of miracles—at least those since the early church—also placed him among the secularly oriented philosophes. Moreover, his work of 1690, on pseudonymous literature, probed the disguises of authors and their multiple motivations for going masked. In other words, the learned Baillet examined his sources carefully; he was entirely familiar with dissimulation and its purposes; and when compelled, he might employ it himself by omitting or merely gesturing at difficulties. As far as we can tell, however, he did not invent.[64]

Yet moderns have treated Baillet's biography with suspicion. From the perspective of the French republics, Baillet seemed too much a Catholic apologist, while in an age of positivism much of the information he reported had to be doubted since it could not be checked against extant evidence. His work came to be superseded by another comprehensive study, Charles Adam's *Vie et Oeuvres de Descartes: Étude Historique*

(*Life and Works of Descartes: A Historical Study*) of 1910, which applied the latest critical methods to the surviving sources. Adam's biography arrived as the final installment of a painstaking twelve-volume collection and annotation of the Descartes materials which he and Paul Tannery published jointly between 1897 and 1910. In addition to reprinting all Descartes's published works, they produced annotated copies of his letters, scraps of evidence about his lost workbooks and other writings, materials about him found in the papers of other people, and closely related primary sources. "Adam and Tannery" has become the baseline for all subsequent studies of Descartes, often showing up in footnotes simply as "AT." Adam's *Vie* certainly made excellent use the material he and Tannery had edited, and he drew on further archival details discovered by antiquaries (such as Descartes's signatures as a witness to family baptisms), providing a verifiable account of his life. It remains indispensable. But while much new information was added, and errors in Baillet were identified and corrected, other events reported by Baillet dropped out because they could not be confirmed. Most important, the hints of aristocratic entanglements and dissimulation were downplayed in favor of seeing Descartes as an upper bourgeois individualist who had always intended to become a philosopher, a hero of modernizing France in the run-up to another great conflict with Germany.

Descartes's religious orthodoxy also seemed to be confirmed by the multivolume publication of the correspondence of Marin Mersenne, a Minim friar and one of the chief organizers of the campaign against unorthodox and even atheistical implications of the philosophy of the period.[65] During Descartes's time abroad, Mersenne was certainly important for keeping him in touch with the views of other savants in France and elsewhere in Europe. Taken at face value, Descartes's relationship with Mersenne seems to place him among the apologists for the Catholic Church. But Mersenne was also corresponding with avowed materialists such as Thomas Hobbes, and when Descartes happened to be in Paris at the time of Mersenne's death on September 1, 1648, he departed without paying his respects to the deceased. We should be cautious, therefore, about assuming that Descartes's "friendship" with Mersenne—which was of benefit to both as an epistolary relationship—implies like-mindedness, much less agreement with positions that later became orthodox doctrines. Even more important, their correspondence cannot illuminate Descartes's early life, as it only begins in October 1629. As a matter of fact, any letters to or from Descartes before that year—that is, during Descartes's first three decades and more—are scarce, amounting to only about a dozen (and half of these are part of the correspondence with Isaac Beeckman, whom we will meet in due course).[66]

In other words, reading Descartes through the Mersenne correspondence and AT's volumes came to mean that the published works continued to dominate the life as the true Descartes. In fact, Adam's *Vie et Oeuvres des Descartes* was not even reproduced in some later reprints of the AT volumes.[67] His life came to be read through his literary persona rather than the other way around.

In Search of a Person behind the Words

During the twentieth century, then, the view of scholars was focused on the mountaintop and what could be seen from there, rather than on the geology from which it emerged. For those concerned with intellectual heights, a stripped-down version of Descartes's life was all that was needed. This was entirely in keeping with twentieth-century modernism, which worshipped at the shrine of high theory. From Einstein's relativity through quantum mechanics and string theory, the ability to reason about the fundamentals of nature abstractly was often taken to epitomize the highest reaches of human thought; other natural sciences also brought their richest offerings to the temple of fundamental concepts, while even in linguistics, philosophy, literature, anthropology, sociology, economics, and many other subject areas—history, architecture, and the fine arts included—unifying theories were given pride of place. For many who visited the temple of theory, "The chief girder in this framework of Modernity" was "Cartesianism,"[68] held to be a dualistic philosophy posing a dichotomy between three-dimensional Nature and unconfined Reason, with preference for the latter. As for the foothills, a consensus arose from a straightforward reading of Adam's biography: after a few lost years of wandering, the maturing Descartes determined on writing philosophy, went off to The Netherlands to find peace and quiet, and despite criticism won through. The incidents of his life were diversions or burdens, pulling him away from the high intellectual aims to which he had always remained loyal. In a kind of confirmation of the mind–body distinction, Descartes's body was necessary only for carrying around his mind, which concentrated on the important work of thinking.

This modernist version of his life has remained powerful. For instance, it frames a handbook of the early twenty-first century written by some of the most learned authorities on Cartesianism.[69] They write that Descartes set out for Germany to sign up with Catholic forces in the opening stages of the Thirty Years' War but that he had no experience of "any armed combat"; he returned to Paris in 1621 after "having definitely abandoned his military career"; his subsequent travels in Italy were required by unspecified "other financial matters"; but at last Cardinal Bérulle "encouraged him to develop his philosophy as an antidote to atheism," which caused him to leave bustling Paris for calm Amsterdam, a place where he could concentrate on writing.[70] The authoritative study of his life by Rodis-Lewis, first published in 1995, elaborates but does not challenge this accepted view.[71] The chief move in recent years—exemplified in the careful and comprehensive intellectual biography of Stephen Gaukroger—has been to question whether Descartes was really the canonical epistemologist or metaphysician, with a growing group arguing that he was more concerned with the natural sciences.[72] The emphasis remains on the development of his ideas, with his life's course mainly providing the backdrop rather than the motivation.

But one also senses a growing frustration with the limitations of the biographical consensus. The philosopher Stephen Toulmin, for example, did his best to give Descartes flesh and blood by investigating his possible responses to the assassination of King Henri IV in 1610. The teenage Descartes was then attending the Jesuit school at La Flèche, where the king's heart came to be interred with great ceremony. Following Adam, Toulmin thinks that the youngster took part in the ceremonies and may have written an anonymous Latin poem for the occasion that compared Henri to the moons of Jupiter, which had themselves just been discovered a few months earlier by Galileo.[73] Toulmin used such examples of the personal and local to press the case against scientific modernism in favor of a more authentic cosmopolitanism rooted in a humanistic openness to experience.

Two other philosophers, Richard Watson and A. C. Grayling, writing biographical studies with a light touch suitable for a general audience, raised further questions about Descartes's motivations and the underlying nature of his real views. In doing so, both came to the conclusion that the usual account could not add up.

Grayling decided that Descartes was a spy. A well-known British philosopher and public intellectual, Grayling was penning a summary of recent work on Descartes, but as he wrote, he must have become increasingly irritated with not being able to explain what was motivating Descartes's travels. So he speculated. When Descartes returned to The Netherlands in 1629, for instance, he did so as "a spy." Grayling explained, "My suggestion is that he was in some way engaged in intelligence activities or secret work during the [previous] period of his military service and travels. It was because of this, I further suggest, that Cardinal Berulle warned him that he was no longer welcome in France." (This suggestion implies that he was both a spy and an exile, which might also make him a double agent.) But Grayling then backs off, saying that he is not asserting that Descartes was a spy but is "merely mooting the possibility," contending that "it is a plausible hypothesis, and merits its place in his tale."[74] Yet later he suggests it as the reason Descartes first went to the Netherlands in 1618 as well, going to the country during the Synod of Dort and leaving it when it was settled, implying that "he could have been sent as a pair of eyes and ears to observe how matters stood in the Breda garrison of Maurice's army while the Arminian difficulties were going on."[75]

Now, great personages of the day certainly did employ spies. In the English-speaking world of the period, the most famous spymaster had been Queen Elizabeth's servant, Francis Walsingham, but in Descartes's own generation the French Cardinal Richelieu became equally famous as a master intelligencer, with eyes and ears in every corner.[76] From a few years later, the powerful minister Jean-Baptiste Colbert has rightly been termed an "information master."[77] Then as now, powerful people desired access to information about their friends and enemies, with both personal and political decision making requiring the obtaining and retaining of secrets. Indeed, the "secretaries" of church and state were so called because they had access to "secrets." Such men were certainly not above applying the methods of both fear and reward in recruiting informants. There are also many well-documented cases of scholars serving as intelligence gatherers and go-betweens, people such Sir Theodore de Mayerne, physician to Henri IV of France and James I and Charles I of England; or the martyred Giordano Bruno; or the German virtuoso Johannes Becher; or the Dutch savant and experimentalist Nicolas Hartsoeker; or later still, the famous Voltaire. There are even suspicions about why Baruch Spinoza paid a long call on French officers after they had taken the city of Utrecht in the invasion of 1672.[78] Such well-known persons might be thought to be more like diplomats than undercover informants, but they certainly collected and passed on private information in secret.[79]

In Descartes's case, there are hints. For instance, we can turn our attention to events following the Peace of Westphalia in 1648, in which France's ally, the Dutch Republic, withdrew from the war against the Habsburgs, leaving France to continue the fight on its own. As Watson points out, it was shortly after the treaty—which also removed from the conflict another French ally, Sweden—that French ambassador Chanut began in earnest to attract Descartes to Stockholm.[80] At the time, one of Descartes's most loyal friends in the Dutch Republic was Henri Brasset, then resident to the French embassy in The Hague. Brasset wrote to Descartes about how he placed some hope in the young new stadholder of the Dutch Republic, William II, Prince of Orange.[81] William would indeed open secret negotiations with France about seizing the government of the Dutch Republic and then dividing the Spanish Netherlands between the kingdom and the Dutch Republic: following William's march on Amsterdam in 1650, only his death from smallpox saved the republican system of the United Provinces from being turned into a principality.[82] Given Brasset's surviving message, the possibility that the young Descartes was an intelligence gatherer, or a well-positioned intermediary, must be taken seriously.[83]

If he was, however, we want to know for whom he was working. Grayling infers that "if Descartes was an agent of some kind, he was by far most probably so in the Jesuit interest," and Jesuits, he further asserts, were agents of the Habsburgs.[84] But this inference rests only on the well-known facts that Descartes had been educated at a Jesuit school and later hoped that his work would be acceptable to the order (it was not). It is also a far too simple characterization of the political position of the Jesuits during Descartes's lifetime. Besides, the Jesuits were spread all over Europe and picked up information everywhere, so there is no need to suppose they required spies who were not members of the society in order to acquire news of major public events such as the Synod of Dort, as Grayling supposes. As Grayling himself notes, too, there is an additional problem, for "if Descartes was an agent in the Jesuit-Habsburg interest, he could not have been conformable with the policy adopted by his own country."[85] French interests were better expressed by Brasset's implied notion that the French and Dutch might seize and divide between them the Habsburg territories on their borders. In other words, following Grayling's hypothesis, Descartes would have been a traitor.

The American, Watson, on the other hand, wants Descartes to be a good guy—that is, in his view, working for the Protestants. He expressed his anger at the "Saint Descartes Protection Society" that made him into a good Catholic, terming his own work a "skeptical biography, as full of doubt about tradition and authority as was Descartes himself," although he also offered plenty of speculation as well.[86] He wrote, for instance, that Descartes's return to The Netherlands in 1628 "was a revolutionary political act." It coincided with the defeat of the Huguenot stronghold of La Rochelle and the flight of many leaders of the Protestant French to The Netherlands (he claims). "It is as plain as can be," Watson concludes, "that Descartes's move . . . was an act of solidarity with republican French Protestantism against royalist Catholic totalitarian oppression, and of liberal Christianity against the Spanish Inquisition."[87] He also thinks that the story of the friendly interview with Cardinal Bérulle—whom he libelously terms a "genocidal maniac (and I speak precisely)"—was invented by Descartes's first biographers, but that if it did occur, Descartes was being privately warned about his own position and so "may have seen flight as the only way out," which would explain why Descartes's subsequent secrecy was important: he "certainly did seem to feel threatened."[88]

Such moves toward trying to understand Descartes's life in light of the politicoreligious situation of his day are important and right, even if they are wrong in their speculative conclusions. If alert to them, one can notice such attempts in earlier biographical studies, too. A century ago, for instance, Adam noted some of the ways that Descartes did not always fit with expectations, wondering at the "double game" that he seems to have played.[89] This line of questioning owes an even greater debt to Gustav Cohen, founder of the forerunner of the Institut Français in Amsterdam, who published a study of French literary travelers to the Dutch Republic in the first half of the seventeenth century, among whom was Descartes. Cohen's main theme was that liberty of conscience in Holland, coupled with the political alliance of 1598 to 1648, drew many French visitors northward, including the soldier-poet Jean de Schelandre and the powerful literary duo of Guez de Balzac and Théophile de Viau, and finally Descartes himself.[90] Guez de Balzac became a close early friend of Descartes, while the one passage of poetry quoted from memory by Descartes in his surviving letters is from Théophile.[91] But Théophile was also one of the most notorious French libertines of the period, burned in effigy in front of Notre Dame de Paris. In the course of treating Descartes as a visitor and resident in the low countries like Balzac and Théophile, Cohen uncovered new information not known to Adam and Tannery and, more important, stumbled across a host of personal and intellectual associations that raised further questions about Descartes's religious orthodoxy. Accounts such as Maxime Leroy's *Descartes: Le Philosophe au Masque*, as well as swashbuckling literary tales such as Dimitri Davidenko's *Descartes le scandalleux*, followed up with other alternatives to the common narrative.[92]

Intellectual historians similarly wondered. René Pintard, the historian of the freethinking *libertins érudits* (philosophical freethinkers), puzzled at Descartes's closest Parisian friends being either wanton gluttons and debauchees or alchemists and astrologers.[93] More recently, after years of relative neglect, the importance of libertine philosophy of the period has been studied again, continuing to furrow brows in puzzlement.[94] The libertines questioned received opinion about Christian moral philosophy, which they took to be rooted in doubtful doctrines derived from disputes about religious speculations rather than founded on the natural sources of human thought and action. On those bases they allowed their own critical lines of reasoning to go to places that made established opinion upset and often angry. One of the first scholars in recent years to raise new questions about French libertine authors was Françoise Charles-Daubert, who in building on the studies of Pintard thought it strange that Cartesians dealt with the same subjects as the *libertins érudits* when they and Descartes seem to have taken no notice of one another in their writings. That was doubly astonishing when it is noticed that they moved in the same circles, at least when Descartes was in Paris. The reciprocal isolation and silence was, she thought, intriguing.[95] It would only make sense if the two groups were fellow travelers but with different and distinct audiences.

Following Charles-Daubert's line, other intellectual historians have noticed more associations between Descartes and the libertines. Catherine Wilson wondered about whether Descartes might have been dissimulating his religious opinions in some of his writings, while Susanna Åkerman began to ask about his association with the mystical libertine Queen Christina; Anne Staquet and Alexandra Torero-Ibad more directly drew parallels between his thought and that of such people as Gassendi and Cyrano, who were previously thought to hold opinions quite other than Descartes's.[96] The views Descartes later stated can indeed be interpreted as important modifications, rather than rejections, of the Epicurean philosophy conveyed to Europe from ancient sources, most notably Lucretius. Recently, the prospect of approaching his works contextually has begun to shape books on Descartes written for English-speaking audiences, as well.[97] But there has been a continued reluctance to explore how the aims of his writing might have been intertwined with the French libertines in his life before his exile.

A study that makes good use of what Natalie Davis has termed "tangential evidence" therefore seems timely.[98] Descartes certainly had friends and adversaries who can be identified, and their views and associations can be examined for possible connections to events in his life. For instance, because Descartes had aristocratic pretensions, he seems to have sought preferment among the royal and noble courts during his first extended stay in Paris. That would place him in the world of the queen regent, Marie de Medici, and a group of nobles who would eventually be driven out of favor. When in 1617 Marie's chief minister and favorite, Concino Concini, was assassinated in a palace coup, Descartes left Paris for the Dutch Republic to learn the art of war. That meant acquiring the skills and knowledge of military engineering, a likely source for the development of his mathematical talents, which would be on display so prominently in later years. The young aristocrat also went on to become caught up in the early events of the Thirty Years' War, and then in the actions prior to the French fighting in the Valtelline in northern Italy; that activity was followed by service in the king's forces at the siege of La Rochelle. His return to France, however, coincided with the rise of Marie's younger son, Gaston d'Orléans, who had close ties with many of the discontented aristocrats. At the time of Descartes's second departure for The Netherlands, at the beginning of 1629, the discontented were being forced into obedience to Cardinal Richelieu or into exile; Descartes left Paris just when Richelieu's agents pushed one of his close friends, Balzac, into banishment to the countryside. Marie de Medici herself resisted but would flee abroad less than two years later, only to die in poverty. Descartes would not set foot in France again until after the death of Louis XIII and Richelieu. Politically, he may even have favored the mixed French and imperial interests of the house of Lorraine, itself a thorn in the side of the French king until Richelieu destroyed their lands in 1635. His personal loyalties were probably aligned with the ideals of his youth, when Marie de Medici sat on the throne.

Abroad, Descartes would begin to write philosophy, no doubt with purpose. The philosophy he would write in the Dutch Republic remained rooted in ideals shaped both by hope for the reconciliation of Christendom and a love of neopagan literature. On the basis of understanding movement in the material fabric of the world, the confessional differences of recent history might be overcome. Descartes may have hoped, too, like many around him, for peace in France and in Europe under the leadership of a loving and charismatic monarch who would be the first among equals in a republic of princes and nobles. Rulers including Catherine de Medici, Henri IV, and Marie de Medici had looked for common interests among the rival factions in the realm and negotiated settlements among them based on respect for lineage and a need for coexistence, while at the same time building alliances in Europe. Their methods, however, were quite distinct from the unforgiving royal absolutism of Louis XIII as fostered by Richelieu, who engaged in foreign wars and destroyed people who hinted at opposition, or drove them away.

Descartes's hopes depended, then, on a renewal at the top of the kingdom of a universal, loving, and forgiving faith that was rooted in natural law. Like many other reformers of his day, Descartes was trying to change the conversation among the autocrats, overcoming division by seeking demonstrable universalities in nature and the laws of nature. The famous argument of the *cogito* was meant as a proof against radical doubt, a reassurance of our own existence from which even ordinary attributes of the real world can be known with reasonable certainty. After all, there were people in his day—as there are in our own—who prefer that nothing can be known for certain. But Descartes considered radical doubt puerile or mad, supportive of authoritarianism. He also hated arguments that had no basis in clear and distinct ideas about the material world and its motions. He consequently came to be seen as a herald of Enlightenment Reason.

*

The mountain in view, then, would seem not to be made from Descartes's own works so much as from the titanic struggles of his moment, layered on top of even more ancient rock. The underlying geology shifts for many reasons, such as the tectonics of struggles for justice and well-being, security and dominion, and flows of capital. But surface events leave their mark, too. The crag is sometimes identified with his name because he came to be such a good guide to it. From personal experience he well knew the pitfalls and crevices to be avoided on the way up, the places where avalanches threatened, and the exposed rock where few handholds could be found. Having scouted out firm paths, and bringing along the latest pieces of kit for the most difficult passages, he could safely and securely bring parties to the top and down again. But he also observed that the people he was guiding enjoyed speaking about their personal experiences, often more real to them than the views in sight. They were on the mountain because they had been moved to come. What that moving spirit was is not for a guide to say, but every guide will understand that visitors are swayed by it as much as the vistas that arise from the journey itself.

Because the mountain now often bears Descartes's name, casual visitors continue to confuse his personal identity with the place. People who live near mountains well know the difference, though. Mountains are not people, despite their moods. Nature on a mountainside is seldom kind, and wildernesses attract not only fierce animals but heartless bandits. To survive there requires not only physical competence but a strong will, a quick wit, and an appreciation for local knowledge. The young Descartes had all these, coupled with a desire to make sense of the larger landscape. And while possessed of active survival instincts, he also wanted to be of service. Before we make our own ascent, then, let's simply stick close by and see what he does.

PART 2

A France of Broken Families

While later in life Descartes became known as a philosopher, he never took a position as a teacher or professor. He came to be well educated and well read and had many friends with whom he could share his keen interest in the latest ideas, but like most other well-heeled young men and women of his time, the young Descartes was often eager to be seen and heard in great houses. He first attempted to rise among the nobility of the sword, although he never quite succeeded. Perhaps he had to make especially energetic efforts because of not having quite the right kind of ancestors in his immediate family, his father being a high-ranking administrator but not of the sword-carrying kind. Descartes was also a kind of orphan, taken in by his mother's family after she died during his infancy and his father remarried. He seems to have broken altogether with his father when they ended up on opposite sides of a great political conflict associated with the rise of Cardinal Richelieu as the king's chief adviser.

But he had a chance. Descartes moved in a world that made judgments according to lineage, title, and office, and both sides of his family descended from royal officials who were addressed by rank—lower-ranking noble titles but noble nonetheless—while the family name itself came from an estate obtained by an ancestor as a reward for military service.[1] He shared his name with the Great René and would seek an equally elevated place, looking for patronage among the grand nobles, strapping on a sword and learning the art of war and international diplomacy. When he came of age, he also gained an inheritance from his mother, which gave him a title of his own. He therefore often introduced himself as the sieur du Perron: lord of Perron. He came from those who valued merit as well, being sent to excellent schools, including law school, giving him further qualifications for high government service. At the end of his life, the inscription on the tomb that was erected over his grave told the onlooker that he was lord of the manor of Perron and descended from ancient and noble lines of Poitou and Brittany; a funeral oration in his honor reiterated the distinctions, calling him a French noble and lord of Perron.[2] When Gerard Edelinck engraved Descartes's portrait for the 1691 biography about him (shown in figure 1 in part 1), he surrounded his distinguished subject with the words *Chevalier* (a noble title) and *Seigneur du Perron*. For added emphasis, below his portrait a coat of arms is prominently displayed, bearing a saltire cross (sometimes also called a Saint Andrew's cross) and four palm fronds.[3] Perhaps the saltire cross echoes the cross of the Valois dukes of Burgundy, since he would later have dealings with members of that august family and places once ruled by them, including Lorraine and the low countries.

The sieur du Perron therefore had sufficient standing to walk among those gathered at princely courts and distinguished assemblies, sometimes gaining the notice of those closer to the center. It was rumored that he was good at the gaming tables, too. Once, he said, he accepted a coin for service as a soldier. As far as is known, however, he took nothing more. Other handsome young chevaliers had risen to favor by serving the great nobles. Playing that game could be dangerous, however, for anyone without personal power walked a tightrope. An awkward gesture might easily cause a fall, and then the cord could be used just as well for throttling. René escaped into exile, where he would begin to write. That Descartes has come to be known as a "philosopher" is perhaps best explained, then, by examining the accidents of his life that foreclosed other options.

Families

When Descartes came into the world, he was granted all the advantages of his family.[4] But he quickly became almost an orphan. The adult Descartes had no clear memory of his mother, only a longing, for Jeanne Brochard died following childbirth when René was only a little over thirteen months old. He and his older siblings, Pierre and Jeanne, seem to have been raised in the house where René was born, under the roof of his grandmother. His father, Joachim, remarried and moved on. Understandably, the youngster would come to identify strongly with his mother's side of the family and as a young adult would distance himself from his father. When he received news of Joachim's death in October 1640, René did not return to pay his respects. His feelings of abandonment are probably captured by the humble arrangements he made for his own burial: placed in an orphans' cemetery outside the city walls of Stockholm.[5]

But although Joachim Descartes may have been an aloof parent, he had a distinguished lineage and undoubtedly knew some of the leaders of France. He traced his family pedigree back into the fourteenth century to people near the Valois king Charles the Wise. Moreover, Joachim's great-grandfather had been the Great René, a military figure who acquired, among other possessions, a tract of land called Cartes in the commune of Ormes St.-Martin near Vienne (south of Lyons). The Great René's grandson, Pierre, Joachim's father, became a distinguished physician but held on to several fiefs, calling himself Pierre "Descartes."[6] Pierre's son, Joachim, turned to law and rose high in one of the chief decision-making bodies of France. By the time of his son's birth on March 31, 1596, Joachim Descartes had been a member (*conseiller*) of the parliament (*parlement*) of the powerful province of Brittany for more than a decade. The first requirement of government is to be just, and parliaments everywhere in Europe were formal bodies deciding on whether the law, or cases at law, accorded with justice. The French parliaments of the time were more like courts than representative assemblies—that better characterized the occasional Estates-General, which last met in 1614–15. (The young René would visit it, too, in the company of an uncle who sat in it as a representative.) But among the matters on which the French *parlements* decided were not only questions of taxation, administration, and justice, but also the edicts of the monarch. Over the course of time, Joachim would rise to become the most senior officer in the grand parliament of Brittany, the doyen.

Joachim's status was bolstered by marrying into other well-connected families. His first wife, Jeanne Brochard—René's mother—came from a family headed by a distinguished official of the ancient and powerful city of Poitiers (*lieutenant general de présidial*). When Joachim remarried three years after Jeanne's death, his second wife, Anne, was also the daughter of a senior government official, Jean-Baptiste Morin, first president of the Chambre des Comtes in Nantes, an important city in Brittany. The new couple soon settled in the administrative capital of that province, Rennes, and raised three boys and a girl of their own.

In other words, Joachim not only came from a distinguished line but also acted as a high-ranking and respected member of the decision-making administration of the kingdom. He was well connected to other people like him in the region and worked closely with the great nobles who served the king directly, someone more than a civil servant but less than a minister. Men like him were among the officeholders who pressed hard to be openly recognized as men of distinguished position (*qualité*).[7] Put another way, he was, according to distinctions that were beginning to be formalized in his lifetime, a noble of the "robe" rather than of the "sword." Some years later, in 1668, when such things were being regularized by the monarchy, the Descartes family obtained letters of chivalry confirming its noble rank; presumably it was this that allowed the engraver Edelinck to depict René as a chevalier.[8]

It is even possible that Joachim Descartes had served His Majesty

Henri IV in person, taking a part in the unraveling of a spy ring and reporting about it to the king. In 1604, when the incident took place, the personal secretary to the French ambassador in Spain was one "Monsieur Descartes." As told by the king's chief minister, the duc de Sully, the Descartes in the story has no first name, but at the time there are no persons other than Joachim Descartes known by that surname. Sully was long the royal governor of Poitou, where the elder Descartes lived at the time, so he may well have remembered the man's last name. Joachim is likely to be the person mentioned.[9]

The incident was this: early in the century, French ambassadors had begun to notice that decisions of the king's council were being conveyed to foreign powers even before they themselves had word. It was later discovered that a clerk who worked for the chief diplomat on the king's council, the marquis de Villeroy, was spying for the Spaniards. That clerk, Nicolas L'Hote, was uncovered by a turncoat, John de Leyré: De Leyré had gone into exile in Spain after plotting against Henri IV in the civil wars, but he now wished to return to his homeland and so shopped L'Hote to the ambassador in return for a pardon and a pension. When the ambassador sent the news to Paris, though, it traveled by networks familiar to L'Hote, enabling him to learn that he had been discovered, which he hastily brought to the notice of his Spanish contacts; De Leyré in turn quickly found out that he had been compromised and, accompanied by Descartes, he jumped on a horse to make for the French border at a gallop to avoid the Spanish authorities. (One can infer from this account that Descartes had been De Leyré's handler.) They successfully crossed the border and made their way to the French court, reporting events directly to His Majesty, and Descartes followed up with a personal report to the chief minister, Sully, as well. Hearing that the game was up, L'Hote tried to escape from Paris north to the Spanish Netherlands, but with pursuers hot on his heels, he drowned while trying to swim the Marne.[10] If the Descartes in this drama was, as is likely, René's father, Joachim, then he was personally known to, and trusted by, the chief persons of the kingdom.

Joachim was therefore an influential man, and he was not averse to pulling strings to help his children. René's elder brother, Pierre, also later became a member of the *parlement* in Brittany,[11] and Joachim even obtained letters from the king himself allowing another of his sons, Joachim II, to succeed him in his office in return for services rendered to the king over the past forty years.[12] No doubt there were other occasions when his trusted services were called on, one of which we will soon have occasion to recount. Whether he tried to do very much for René, however, is doubtful, since the youngest son of his first wife chose not to follow in his footsteps.

René had, after all, mainly been raised by his mother's family. In his infancy and early childhood, Descartes seems to have grown up in the home of his maternal grandmother, who had moved to a pleasant town on the river Creuze. Grandmother Jeanne Sain seems to have been a strong-minded woman who had moved to La Haye after separating from her husband years before. René's mother came to her mother's house when childbirth was imminent; there Descartes was born and baptized (in the Catholic rather than the Huguenot church); and there his mother died a year later. The little children were probably living at their grandmother's when their mother left the earth, and Jeanne Sain probably simply continued to keep them with her and her servants. Perhaps being raised in a household whose head was female helped cause Descartes to listen carefully to women in later years.

As he grew older, Descartes probably also spent time at the home of his paternal granduncle and godfather, Michel Ferrand, in the nearby city of Châtellerault. As the second city of Poitou, it would have had first-rate Latin schools for his early education. Ferrand was an important figure, serving as councilor to the king (a parliamentary office) and lieutenant general of his city (a royal appointment), in a most difficult time. Like much of the rest of Poitou, Châtellerault had come to be more or less equally divided between Protestant and Catholic, and Ferrand was probably involved with passing the local ordinance in 1589 that allowed Protestants the free use of their religion, making the city one of their "villee de sûreté" (safe towns). In any case, as lieutenant general, he oversaw the local negotiations for the Edict of Nantes in 1598 that gave the Huguenots full civil liberties. The edict was a fundamental component of King Henri IV's strategy to forge civil peace after the wars of religion, a political line continued by his widow, Marie de Medici. (Châtellerault would not send a representative of the Catholic clergy to the Estates-General in 1614, probably indicating its continued religious balance.)[13] Ferrand had also been associated with the unorthodox but ecumenical political theorist Jean Bodin.[14] Interestingly, too, in the house next door lived Ferrand's cousin, Gaspard d'Auvergne, who had served as an ambassador from King François I in Italy and translated Niccolò Machiavelli's *Prince* and *Discours* into French, two very important works that many commentators termed atheistical but which Descartes later wrote about very knowledgeably.[15] Descartes must have grown up with the latest political discussion as a commonplace.

It would appear that his mother's family saw to his further education. In 1606 or 1607 René and his elder brother were sent to the Jesuit college at La Flèche. The Jesuits had been accused of complicity in an attempted assassination of Henri IV in 1594 and were expelled from the kingdom, but following its readmission to France in 1603, the Society of Jesus organized several fine advanced schools for the sons of nobles and gentlemen. For the school that René attended, Henri gave them the château in which he had been conceived, making it known as the "Collège Royale."[16] The rector of the school (Father Charlet) was allied with his mother's family. Descartes later referred to him as his "second father."[17] Descartes apparently left school about the age of sixteen—entirely in keeping with the customs of the time—for he spent the winter of 1612–13 at his father's house in Rennes, training in horsemanship and fencing.[18] But his mother's relatives again seem to have organized his further education, at the university in Poitiers, a city where his maternal uncle and godfather, René Brochard, sieur des Fontaines, held the office of councilor to the king. When Descartes took his degree and diploma in law from there in 1616, he dedicated his thesis to Brochard, who also paid his fees.[19] Descartes would inherit property in Poitiers from his mother's estate, too. While all other members of his family claimed association with his father's Brittany, then, René always identified himself as from the province of Poitiers and Châtellerault: his mother's Poitou.[20]

Politiques

In the autumn of 1614 and beginning of 1615, his uncle Brochard and the eighteen-year-old René were in Paris for the important assembly of the Estates-General, where other relatives also sat.[21] The Estates-General brought together representatives from throughout the kingdom to help the queen regent, Marie de Medici, settle the realm after a failed insurrection by a group of discontented nobles and princes of the blood. Marie had been particularly concerned with the loyalty of Orléanais—the region represented in part by Descartes's relatives—but in the end her government managed to find trusted delegates from there.[22] Poitiers was particularly deeply split between the politicians and the bishop, who took control of the city in the name of the crown. Poitiers even shut out the malcontented Prince de Condé, who retired to Châtellerault, presumably because that city remained loyal to him.[23] Uncle Brochard must have been a remarkably able politician in such circumstances, for when the Estates were called to order he sat for Poitiers (giving his title as *ecuyer*, or squire, the lowest rank of nobility). A representative for adjoining Touraine was a cousin: Maître René Sain, councilor of the king, treasurer general of France, and mayor of Tours.[24] Another cousin, from Descartes's father's side, Maître François Ferrand, councilor and lawyer of the king, represented Châtellerrault.[25]

As representatives of the third estate—the *peuple* (people)—Descartes's relatives were among those calling for a stronger but reformed monarchy, along lines that today would be termed secular. Their position was a further move in the loyalist *politique* program that had emerged in the 1560s and remained important in France during Descartes's youth. The *politiques* spoke out in response to the horrors of the

French religious wars, arguing that good government should be placed ahead of religious doctrine or other ideologies. Great authors such as Bodin and Montaigne gave it the voice of toleration. Henri IV put it into action by abjuring his Huguenot faith in favor of France's civic religion, Catholicism, while at the same time allowing others to practice their Protestant faith openly as long as they remained loyal to the state. In the Edict of Nantes, Henri gave the political sphere autonomy, for even while sacralizing the person of the king, the edict did not identify the monarchy with Catholicism alone.[26] People such as Jacques Auguste de Thou, a senior member of the *parlement* of Paris and a royal official, took to print to argue that the rule of law should reign supreme over even the greatest persons.[27]

The third estate was packed with people who shared the outlook of De Thou, including Descartes's relatives. Their chief political position was to advocate something even stronger than the *politique* position, something which has been called "political Gallicanism": to make the office of King of France sovereign over all persons in the kingdom, whether noble or clerical. The Orléanais, which included Brochard, were among the most vehement advocates of that line.[28] (It turned out to be a position stronger than the crown itself wanted.) One can easily imagine that Descartes, too, had come to be sympathetic to this legalist and nonclerical view of sovereignty: he would later argue that even God was not greater than his decrees.[29]

The strong association of *politique* views with Henri IV must have already made an impression on Descartes at school, especially during the commotion following the king's assassination. Many writers commenting on Descartes's early life have drawn attention to the likely influence on him of that powerful event. It occurred in 1610, when Henri was raising an army in preparation for helping the elector of Brandenburg in a war with the Habsburgs over the succession to the duchies of Cleves-Jülich, which occupied a strategic position on the Rhine. Before Henri's departure for the Rhineland, as a precaution against a return to civil war should anything go amiss, the king agreed to have his wife, Marie de

Medici, crowned as queen of France. The following day he was in a coach with companions traveling through streets in Paris still crowded from the celebrations when congestion forced a halt; a right-wing Catholic fanatic managed to mount the vehicle and stab the king to death. Needless to say, the event provoked shock, anger, and grief throughout the kingdom, along with an extended period of mourning.

The aftermath touched Descartes directly, since Henri's embalmed heart was sent to La Flèche to be ceremoniously laid in a tomb placed high above the alter of his school's chapel. Hercule de Rohan, duc de Montbazon, who had been with Henri and was himself wounded during the assassination, commanded the procession from Paris to La Flèche, and oversaw the sacral entombment of the heart. Some of Descartes's biographers think it likely that René was among the young gentlemen who accompanied the cortege.[30] If so, perhaps it began a personal connection between him and Montbazon: Descartes's birthplace, La Haye, had been acquired as part of the duchy of Montbazon, so it is not improbable that he was recruited for the ceremony as one of Montbazon's subjects studying at the school.[31] The connection seems to have continued, for Montbazon's daughter, Marie de Rohan, best known by her later title, duchesse de Chevreuse, had a son, the duc de Luynes, who later translated Descartes's *Meditations* from Latin into French with Descartes's help. But perhaps one can find further echoes of the event later in his life, too. The sepulcher in which Henri's heart was placed was shaped in the form of a pyramid; so was the tomb raised over Descartes's burial place.[32]

Breaking with His Father

But the political state of France was fraught, sometimes dividing families. Later in life a distinct coldness is evident between Descartes and his father. Perhaps the break came in 1626 over an incident that had much to do with both the personal and political conflicts within the ruling family of France, echoed in the conflicts in other clans. The episode exposes the frictions within the Descartes family, too.

The just-mentioned Marie de Rohan was the chief mover in a grand intrigue. By that time the younger Descartes had—as we will soon see—gained considerable experience as a gentleman volunteer with various armies in northern and central Europe, with a further period in Italy. In 1625 he had returned from his travels, taking up residence in Paris and deciding his next move. There he lived in the home of a friend of the family, Nicolas Le Vasseur, sieur d'Étioles, yet another high-ranking member of the royal government (receiver general of finances). According to an early biographer, Baillet, Descartes was at the time toying with the idea of taking up an important royal post that had come vacant, that of lieutenant general of Châtellerault, a position that his godfather Ferrand had previously held.

By then René was widely traveled and valiant, and he possessed a law degree and friends in government. All he had to do was to come up with the sum of fifty thousand livres necessary for the purchase of the office and presumably secure his father's blessing. The first of these he had, having returned from Italy with thirty thousand livres, on top of which he sold some additional properties that summer, and for the rest he had an offer of a loan from a friend. Most of Descartes's biographers therefore wonder why he did not take the post and settle down to begin a family, as might be expected.[33] But perhaps his father balked. The only recorded comment of the father on his son was a much later snide remark about how he had been "bound in calf-skin," implying that by becoming a writer his son had failed to live up to his expectations.[34] René himself later commented that his father thought he was not experienced enough for the office.[35] It must have hurt. Or perhaps, in light of what followed, his father's opinion was a veiled reference to his son's inability to bend sufficiently with the political winds.

The winds were just then blowing fiercely. In the spring of 1626, the question of the lieutenancy not yet having been decided, Descartes traveled from Paris to Poitiers and Châtellerault together with his relative and landlord, Le Vasseur. Having looked things over, they turned north to Rennes to see his father.[36] As it turned out, however, Joachim Descartes was embroiled in one of the great political events of the moment, a rebellion of many of the great nobles stirred by Marie de Rohan that resulted in the show trial of the comte de Chalais. The trial would be organized by Armand Jean du Plessis, bishop of Luçon and now Cardinal Richelieu. By 1626 Richelieu had become the king's indispensable chief minister, growing into one of the firmest strategists and most supple tacticians of his generation. Previously unnoticed by Descartes's biographers is the fact that one of the chief judges on the court was Descartes's father.[37] If the younger Descartes had friends among those who were opposed to the cardinal—which, as we will see, seems likely—then his father's willingness to do Richelieu's bidding might well have caused the son to separate.

René Descartes would have noticed Richelieu no later than at the Estates-General of 1614, when Bishop du Plessis sat as a representative of the first estate (the clergy) for Descartes's homeland of Poitou and began his climb into the councils of the crown.[38] About a decade older than Descartes, from a somewhat more eminent family and first intended for the life of the sword, Richelieu took over the family bishopric when the elder brother for whom it was intended decided instead to become a monk. Cardinal Richelieu became an active Catholic reformer, following Rome in being the first bishop in France to introduce the decrees of the Council of Trent into his diocese. At the close of the Estates-General he was chosen by the clergy to summarize the first estate's position, powerfully arguing against the third estate's position about the sovereignty of the monarch over the church, favoring the Tridentine decrees despite the crown's firm independence from Rome on this issue, and concluding with personal praise for Marie de Medici.[39] He made a name for himself. Within a year, following the marriage of the Spanish princess Anne of Austria to Marie's son, Louis, he gained the position of Anne's chief almoner and began to advise Marie de Medici and her favorite, Concino Concini, as well.[40]

But in the mid-1620s the royal family was quarreling. Not only had the queen mother, Marie de Medici, and her eldest son, King Louis XIII, been at swords drawn, but more recently her younger son, Gaston d'Orléans (see fig. 4), was known to be waiting in the wings to succeed

Figure 4. Gaston d'Orléans, Monsieur the Heir Apparent, as a military leader. Attributed to Claude Deruet, Château de Blois. Gianni Dagli Orti / The Art Archive at Art Resource, New York.

his often seriously ill and childless brother. By 1626, with Gaston now age eighteen, many of the great nobles saw their interests aligned with "Monsieur" but found that Richelieu stood in their way. At Gaston's death in 1660, his funeral oration was suppressed because of the objections of his nephew Louis XIV, but a summary later found in the archives emphasized that Gaston had worked to "maintain the liberty of the people without damaging the authority of the prince."[41] "The people" indicated the political classes of France, both the discontented nobles and the *peuple* who sat in the third estate, including Descartes's relatives.

As a member of "the people," Descartes was a close observer of various aristocratic types, and what he later wrote might apply to the Gaston–Richelieu conflict. He would comment on how self-esteem is a principal part of wisdom, arising chiefly from generosity of spirit. People who truly possess generosity therefore think well of others, are usually the "most humble," and "are naturally led to do great deeds," while at the same time they never undertake what they cannot perform. Armed with an accurate sense of self-worth, they "are always perfectly courteous, gracious and obliging to everyone." That was the persona of Monsieur. But unfortunately, many of the great fell into the trap of vanity, which "is always a vice." They did so because "flattery is so common everywhere."[42] It made for viciousness. Such was the view of Richelieu, surrounded only by his subservient creatures, held by the discontented. Since remaining humble and generous in the face of constant flattery was no easy task, the vainglorious often pushed themselves forward and forced events to obtain personal advantage, bringing havoc in their wake.

As long as Gaston remained without a wife, many aristocrats entertained hopes of establishing a dynastic alliance with the royal line through marriage, increasing Monsieur's ability to build a large network of clients seeking his favor. Richelieu therefore wanted the prince married off to minimize his power as an object for intrigue. The plan, in keeping with Marie de Medici's desires, was to marry Gaston to the young and wealthy heiress Marie de Bourbon, duchesse de Montpensier; that would also keep the duchesse out of the hands of one of Richelieu's adversaries, Louis de Bourbon, comte de Soissons, who claimed that he and she had been betrothed. There is a possibility that the younger Descartes himself had some opinion about the matter, for among the duchesse's titles was Lord of Châtellerault, presumably giving her some interest in the lieutenant general's post that he was just then considering.

But in response to Richelieu's plan, a group of the great nobles decided that it was finally time to act against the minister, and they plotted his assassination.[43] One of the chief movers in the drama held great estates in Brittany: Marie de Rohan, duchesse de Chevreuse, a distant claimant to the throne herself. Marie de Rohan was not only one of the most brilliant and charismatic women of the time, but she was also the daughter of the eminent Montbazon who had commanded the funeral procession that carried the heart of Henri IV to Descartes's school. Marie was a confidant of both the queen mother, Marie de Medici, and the

Figure 5. Anne of Austria as Minerva, goddess of wisdom, arts, trade, and war. Painted by Philippe de Champaign, Conseil d'Etat, Paris. Erich Lessing / Art Resource, New York.

queen, Anne of Austria (see fig. 5). Not many years earlier, Marie and one of her lovers, Henry Rich, had arranged a private meeting between the Duke of Buckingham and Queen Anne that allowed Buckingham to pledge his love, causing much public consternation. Then, while in England, Marie so scandalized the court by her behavior that Richelieu let it be known that she was a liability, although her marriage to the duc de Chevreuse meant her person could not be touched; she in turn openly rounded on Richelieu. Marie and her conspirators among the aristocrats now saw an opportunity to do away with the cardinal and perhaps even (it was rumored) depose Louis and wed Gaston to the much-neglected Queen Anne, at last also giving the hand of Marie de Bourbon, duchesse de Montpensier, to Soissons and so generally consolidating the power of the greats over the crown. Some of Marie de Rohan's adherents included some of the highest-ranking nobles in the land: the half-brothers of Louis, the duc and the chevalier de Vendôme (the latter of whom was the royal governor of Brittany); the ducs de Nevers, Longueville, Condé, and Rohan; and of course the comte de Soissons.

Another plotter and yet another of Marie's personal conquests, Henri de Talleyrand, the young comte de Chalais and master of the king's wardrobe, joined rather late in the game. As he was drawn in, he grew frightened at the possible consequences, consulting his uncle, a knight of Malta, who insisted that to save his own life, his nephew had to confess all to Richelieu. He did. Richelieu quickly launched further secret investigations and employed imprisonment and torture to unravel the intended coup, consolidating his own grip on power.

After the plot was discovered, Cardinal Richelieu (see fig. 6) gained the complete support of the king. Gaston's former preceptor, the great *maréchal* (marshal) Jean-Baptiste d'Ornano, was arrested on May 3, 1626, and died in the dungeons of Vincennes in the autumn, rumored to have been poisoned; on June 13 the king's half-brothers Vendôme were also arrested and confined to prison, where one of them, Alexandre, would also die (in early 1629). The king and his court traveled to Brittany for the summer to personally oversee the loyalties of a region that might have risen against them, and it soon appeared that they regained control. But Richelieu discovered that Chalais continued to meet in private with Gaston, suggesting that the cabal continued. On July 8 Chalais, too, was arrested and, threatened with torture, made confessions that implicated other plotters, although he later retracted his testimony, saying that he had given it in return for a pardon. The king subsequently

Figure 6. Richelieu, Cardinal of the Catholic Church, wearing the chivalric medal of the Order of the Holy Spirit. After Philippe de Champaigne, Musée du Château de Versailles. Alfredo Dagli Orti / The Art Archive at Art Resource, New York.

awarded Richelieu a permanent guard of fifty men—setting up the rivalries between various armed units at court that Dumas exploits so well in *The Three Musketeers*—while Richelieu took the occasion to force the marriage of Gaston and Montpensier (which took place on August 5, in Nantes). Shortly after Chalais's arrest, Richelieu also established a special tribunal to dispose of the count. It was Richelieu's first use of a kangaroo court, a method that the cardinal would in the future employ from time to time to spread fear among his enemies.

The special tribunal was staffed by members of the *parlement* of Brittany, including Joachim Descartes, who was listed as the third most senior member of the eleven-person court.[44] They were officially called into session in Nantes on August 5 and sat for the first time on August 11. After taking depositions from various conspirators and deposing Chalais himself, they declared his guilt on August 18. He was to be executed the next day. Some of the comte's friends made an attempt to preserve him by smuggling the executioner out of town, but another prisoner was promised his liberty if he would do the deed. Then the executioner's sword disappeared. Another was volunteered from the crowd. But it was apparently not sharp and, coupled with the performance of an amateur, Chalais was left bleeding and screaming in pain, still alive after twenty strokes. His neck was finally struck from his body by repeated hammer blows against a cooper's adze. Queen Anne herself barely avoided imprisonment or worse.[45] Others slipped away. In the case of the duchesse, she fled to the independent duchy of Lorraine, finding refuge in her husband's family.

We do not know what either Joachim or René thought of the business, but it was nothing if not ugly, and it must have touched them both—especially if the distant son had friends among the nobility who were unhappy with the cardinal, which, as we shall see, was likely.

René returned to Paris and did not pursue the government post. One wonders, too, about whether the event contributed to the father's resignation of his seat in the *parlement* a year later in favor of his son, Joachim II.[46] But the elder Joachim soon also went to the trouble of obtaining a further letter of honor from the king.[47]

Aristocratic Paris

It might seem odd that Descartes apparently became loosely associated with the anti-Richelieu faction in the late 1620s. But many great nobles and their followers who were loyal to the monarchy in the 1610s, when Descartes first spent time in Paris, would later oppose the cardinal. In later years, Descartes's *Discourse on the Method* (1637) gives some autobiographical hints about his first encounters with politics, although he is very cautious. His main point is that "as soon as I was old enough to emerge from the control of my teachers, I entirely abandoned the study of letters." Instead, he resolved "to seek no knowledge other than that which could be found in myself or else in the great book of the world." For the rest of his youthful days, then, he spent his time "travelling, visiting courts and armies, mixing with people of diverse temperaments and ranks, gathering various experiences, testing myself in the situation which fortune offered me, and at all times reflecting upon whatever came my way so as derive profit from it."[48] Descartes had been a talented student who had earned the ability to read more or less anything in his school's library,[49] and his excellent preparation allowed him to understand and comment on whatever he wished later in life. But for the years following school he paints a youthful rebellion and the search for experience of the world, together with trying to make sense of it all.

He may have been living in Paris as early as 1613, at the age of seventeen, a common enough age to be sent into the world. Beforehand, in Rennes his father had provided him with further training in riding a warhorse and handling weapons, suggesting that—like Dumas's D'Artagnan—he saw his future among the young noble retainers of one of the great houses.[50] The key word in Descartes's own summary of his earliest activities, then, is *courts*. According to Baillet, Descartes had wished to join the company (*troupes*) around the young king.[51] But Baillet also apologizes retrospectively for the father's fault in allowing his son to reside in the city with only a valet (i.e., without a tutor or chaperone), suggesting that he was investing only the minimum in his distant son's upbringing. Without connections, Descartes would have to work his way into one of the groups around the court by proving his worth. He certainly seems to have found companions. Apparently Descartes was skilled at the gaming tables, and Baillet is clear that the young and handsome Descartes took great pleasure in music and poetry as well. His nights must have been full.

But Baillet also comments that Descartes's companions became concerned about how he frequently absented himself from their divertissements, making them wonder whether he had a secret life or even worrying that he had returned to Brittany or had gone with the court to Guyenne for the exchange of brides between the French and Spanish kings. One absence might be explained by Descartes serving his uncle Brochard during his time in Paris for the Estates-General. But one might also wonder whether Baillet is hinting about how the young cavalier traveled among the great cloud of courtiers to southwestern France for the royal marriage in late 1615 a few months later. It would have been an opportunity to work his way into circles around the young king. But apparently he had impressed his uncle, too, for on the court's return to Paris they traveled through Poitiers early in 1616, where Descartes must have stopped for a few months of further study, since the now twenty-year-old Descartes collected his law degree and license in November, dedicating his thesis to Brochard.[52] According to Baillet, the young man then returned to Paris during the Christmas season of 1616, when his Parisian friends welcomed him "as merrily as they possibly could."[53]

In other words, during his late teens Descartes was most probably acting like others of his kind, busying himself with rounds of visiting and socializing, and perhaps swordplay, looking for the main chance to attach himself to one of the groups around the great persons of the day. He would have been running risks, however. For while the courtly life could be pleasurable for anyone, without the routines of officeholding it could also be exhausting and confrontational. The personal and political were so intertwined as to be indistinguishable at times, as the governing words for proper manners indicated: *savoir-vivre* and *savoir-faire* (ways of knowing how to live well) were put into action by *politesse* (politeness), etymologically related to *politique* (political). Codes of honor and orders of chivalry were held in high esteem. But authority depended not only on place and rank, and inherited networks of familiarity established over generations, but on personal charisma as well. Prince or Princess "Charming" exercised a kind of magical power over the spirits of those nearby. Conversation was a form of intercourse in which emotional and sexual webs could be woven, loyalties tested, and opinions weighed. Speech required wit and discretion, and dissimulation was an art practiced by most, allowing one to say only what was diverting or, if necessary, words that were truthful but suggestive of agreement when there were differences.[54] Power over others could be conveyed by displays of confidence, the studied nonchalance that the Italians called *sprezzatura*, never seeming to worry but giving with an open hand and laughing at danger and threat, always displaying spirit.

Like others of his kind, Descartes carefully observed in those around him any signs of conflict between authentic feeling and calculation. In his last published work, on the passions, he noted how expressions of the eyes indicate every feeling, although it was almost impossible to describe those slight but meaningful movements. Facial expressions were almost as difficult to scrutinize and control. Some, however, could be managed, such as wrinkling of the forehead or movements of the nose and lips, which with practice could be made voluntary. "In general the soul is able to change facial expressions, as well as the expressions of the eyes, by vividly feigning a passion which is contrary to the one it wishes to conceal. Thus we may use such expressions to hide our passions as well as to reveal them." The deep responses, which arise from the seat of the passions, the heart, via the blood—such as going pale or blushing, or trembling—were virtually impossible to control even by the most practiced, but with polish, courtiers could act their parts without revealing themselves.[55]

Even well-placed persons, both men and women, lived with daily frictions. They were surrounded by others from their rising until their retiring, needing to see that subordinates carried out their duties correctly and otherwise working the scene with smiles, scowls, or icy distancing so as to distribute love and fear appropriately. Any perceived slight from an equal had to be confronted immediately and sharply, or else it might become a festering wound assuaged only by innuendo, backstabbing, and poisoned chalices. Given the latter risk, the loyalty of kitchen staff and apothecaries, among others, required constant checks. Peers sometimes also required a slap in the face: hundreds died each year in sword duels, including many of the court favorites. Even a noblewoman, Madame de Saint-Balmont, of Lorraine, became known as an adept at swordplay.[56] Louis XIII attempted to prohibit the settling of affairs of honor in such ways by enacting a law in 1626—and by making an example by executing the powerful noble François de Montmorency-Bouteville in 1627 after he dueled on the grounds of the Palace Royale—but with only limited effect. Whether high-ranking or not, then, one had to be constantly on guard. Various authors have commented on Descartes' thin skin, his quick and vehement responses to any hint of disrespect even from old friends, but in this he was acting like others of his class. Honor and valor demanded it.

There would be daily times for religious observance, too, and for the nobility, at least, further opportunities for hunting, visiting behind doors, engaging in handiwork, and meal taking, together with attendance at weekly concerts and balls. The pleasures of gambling and drinking, and lovemaking, along with lighthearted and witty conversation, certainly focused attention. Those who lived after dark could afford candles and torches, and they decked themselves with jewels that sparkled in reflected light: in the sixteenth century, European jewelers had learned how to cut gems into facets. It is estimated that early in the reign of Louis XIII, the staffing of the royal households (king, queen, and queen mother), with their kitchens, halls, guardrooms, stables and kennels, courtyards, chapels, and rooms public and private employed well over two thousand people.[57] Such numbers do not count those serving in the royal government and armed forces, nor guests and visitors. When in the early 1560s the court set out for a long period of traveling through France, more than ten thousand people accompanied the royal family.[58] The greater nobles had comparable suites, and comparable governments and troops at arms, according to their rank and wealth. Lesser gentlemen and gentlewomen often attended on the greats at public events or in the public rooms in their households, sharing information and rumor with one another, receiving the hospitality of their hosts and hostesses while also trying to earn their recognition and favor.

Any routines were constantly interrupted by important days of the religious calendar and business year, memorial days such as birthdays and saint's days, and even by changes in the weather and public events, whether executions or stage shows. Some events were unplanned, as when epidemic disease or armed conflict devastated a community, but others were occasions for lavish celebration. For instance, in 1612 Marie de Medici decided to commemorate the contracted double marriage negotiated between her family and Spain with magnificent gaieties, including jousts held at the Place Royale. *Lists* (double tracks down which men on horses carrying spears charged at each other) were constructed that stretched more than two hundred and forty feet, surrounded by platforms rising more than 12 feet in height to give spectators (estimated at ten thousand persons) a good view; in the middle stood a pavilion for Their Majesties, draped in the most expensive fabrics. The duc de Guise, the duc de Nevers, and the marquis de Bassompierre organized the action, which was made up of more than five hundred men and two hundred horses, all clothed and caparisoned in silver cloth and scarlet velvet. The jousting continued for three days, from nine in the morning until six in the evening, followed each evening by artillery fire, fireworks, and allegorical processions. It cost each of the sponsors at least fifty thousand crowns: princely sums. The days afterward were occupied by balls, banquets, and tilting at the rings. Then the court moved on to Fontainebleau to receive the Spanish delegation with properly magnificent entertainment there as well.[59]

Only the very great could sustain expenditures of this kind for long, despite the exploitation of their many estates, without resorting to wartime seizure or profiteering. Consequently, the oppression of ordinary people in the period has become famous, and the "population decline" in many parts of seventeenth-century Europe is simply an abstract way of referring to a collective phenomenon that at a personal level was terrifying, with more children and adults dying miserably than infants surviving. What was once called "the crisis of the seventeenth century" has recently been expanded into a vision of a period of worldwide catastrophe.[60] While central Europe would be especially hard-hit by the Thirty Years War, France was traumatized by continued conflict between various nobles and the king, fierce continuing religious warfare, and frequent desperate uprisings.

An example of the consequences of war and disease can be found in a letter from France written by the famous physician and anatomist William Harvey, who late in the summer of 1630 was accompanying one of the English king's young favorites, the Duke of Lennox, from Dieppe to Aubigny before heading on to Paris. In his letter Harvey makes a dark joke about being unable to find even "a dog, crow, kite, Raven, or any bird, or anything to anatomize" in a bleak landscape. Everything had been eaten up. The sentence continues with a chilling remark: here there were "only some few miserable people, the relics of the war & the plague, where famine had made anatomies before I came." Cannibalism? (It was known.) Sadly, "It is scarce credible in so rich, populous & plentiful countries as these were that so much misery, desolation & poverty & famine should [occur] in so short a time as we have seen." He thought that France could not go on in this way, for "it is time to leave fighting when there is nothing to eat, nothing to be kept & gotten," nothing but robbery.[61] Perhaps he exaggerated. But on occasions when Louis XIII was told about the condition of his people, he sometimes fell to weeping. What could he do? The sinews of power, both armies and public display, required coercion and money, his debts were huge, and so his tax farmers had to search under every bed to find every penny. Almost no one lived in ease and comfort, with even the greatest scrambling for the next chance.

In such a world Descartes was unusual in apparently remaining masterless for most of his life. But masterlessness was a rare condition even among lesser nobles, for not serving someone of greater rank or means in person or through officeholding threatened vulnerability and poverty. Descartes's reported absences from his friends may, then, have to do not only with traveling with the court and taking a law degree in Poitiers—which he kept so close that Baillet did not know of it—but with his attempts to get ahead by cultivating possible patrons. We know that about the time of his return to Paris, Descartes also became close to Claude Mydorge; Baillet says that they were "inseparable."[62] Ten years older than Descartes, Mydorge was one of the royal judges (*conseillers*) in the high court of the Grand Châtelet. Mydorge's father held positions in Paris similar to Descartes's father in Rennes, while his wife, Mademoiselle de la Haye—carrying the same name as the town where Descartes was born—was daughter of one of the most senior civil servants of the crown (a member of the court of auditors) and sister of the French ambassador to Constantinople. Mydorge himself later rose higher into the ranks of monarchical government, but he has come down to posterity as one of the foremost mathematicians of the period. One can well imagine that given their similar social backgrounds, the older judge, Mydorge, was acting as a friend and mentor to the newly graduated lawyer. Perhaps he also stimulated a serious early interest in mathematics in his young companion. Descartes's absences from his other friends might therefore be well explained by his attentions to Mydorge. If through such personal patronage he also hoped to gain a position in the administration of the realm, that might explain why he was so discreet with his friends about how he was spending his time.

Or perhaps he was quietly making friends in many places. Baillet writes that this was the time that Descartes met the friar Marin Mersenne, with whom he would correspond frequently in later years. There are problems with this, however: Mersenne had also attended La Flèche but, being eight years older, probably had little to do with the younger student at school. Moreover, Mersenne had joined the Minim Friars in 1611 and between 1614 and 1618 was teaching in Nevers, causing later biographers to be doubtful about his meeting Descartes in Paris (although he could have visited from time to time). There is no reason to assume that they became well acquainted during Descartes's first residence in Paris.

More likely, the period saw the beginning of Descartes's friendship with Guez de Balzac, later known as an important literary figure and a rare early correspondent of his. Balzac also came from the same social rank as Descartes—a family with claims to nobility but more like *gentilhommes* (gentlemen) in daily life—had been educated in a Jesuit college (in Poitiers), and was almost exactly the same age.[63] They also had friends in common. By the end of the summer of 1615, Balzac had returned to Paris from The Netherlands, where he had gone with the poet and playwright Théophile de Viau after meeting Théophile's troupe of players. His experience there may also have been what soon encouraged Descartes to head north, too.

While Mydorge understood officeholding Paris well, Balzac would have been a conduit to even higher reaches of society. Balzac's family was in the service of Jean Louis de Nogaret de La Valette, duc d'Épernon, a powerful nobleman on excellent terms with the queen regent and her allies. D'Épernon is sometimes accused of involvement in a conspiracy leading to Henri IV's assassination, but he was certainly one of Marie de Medici's strongest defenders in its aftermath, when she was striving to be accepted as the queen regent. In 1619 he would protect her further by freeing her from house arrest, making negotiations between her and her son possible. Those negotiations were carried out in Angoulême in the residence of the duc's principal secretary, Balzac's father. The younger Balzac himself entered the duke's service in 1618 and would take up his pen on behalf of d'Épernon to paper over the duc's actions on behalf of the queen mother so as to return him to the good graces of the king.[64] Balzac is therefore likely to have been a helpful intermediary to the greats, introducing Descartes to the kind of circles around the queen regent who might have found a use for a discreet and talented young nobleman.

Libertine Paris

Baillet simply refers to Descartes's friends and their divertissements. But if Descartes was associated with Balzac at the time, as is likely, it would also place the young Descartes in the circle around Marie le Jars de Gournay, Montaigne's literary executor. She inherited the great Montaigne's library and in turn left it to François La Mothe Le Vayer, who was not only an advocate in the *parlement* of Paris but one of the most famous of the *libertins érudits* (erudite libertines) as well.[65] (He in turn married Mydorge's niece.[66]) Gournay herself employed her pen on behalf of Henri IV and Marie de Medici, and she would continue to do so for their successors until her death in 1645 at the age of seventy-nine. On many afternoons some of the most distinguished intellectuals of the period would climb the steep and narrow stairs to her third-floor rooms: the informal male academies that were beginning to appear on the scene often met in their hosts' libraries, whereas the less formally educated women hosted their guests in salons or even in their personal rooms.[67] Gournay was on good terms with Théophile, and also with Balzac, who later engaged in a misogynist attack on one of her rivals, Charlotte des Ursins, vicomtesse d'Auchy (the mistress of the writer François de Malherbe).[68] There are certainly echoes of Montaigne in Descartes's later published works, not only in general ways such as his use of doubt and of the first person singular to explore and explain his world (and his refusal to become engaged in controversies about religious doctrine) but also in a passing reference to the cultures of cannibals and of China in his *Discours* that reminds one of Montaigne's famous comments on the dignity of other cultures.[69] Descartes is known to have been keen on poetry and literature. In later life he also happily accepted women as his intellectual equals. It is very possible, then, that the young Descartes attended Gournay's salon with Balzac.

Descartes's connection with Balzac would also place Descartes among the young libertines of Paris. The word *libertine* meant "freethinker," although for critics it implied a life free from the strictures of any moral rules other than those for which a natural basis could be found: that is, many observers considered libertines irreligious and immoral. Yet both libertine ideas and behavior were common in Paris at the time. Even the model of courtliness, *Amadis of Gaul*, is full of knights bashing one another to pieces, secret potions and dark forests, people who turn out to be other than they appear, and beautiful women happy to indulge in mutual delight with their champions; there are many solemn vows, but there is hardly a priest or cross in sight, and no theological discussion. The fondness in the period for classical literature, particularly Ovid's *Metamorphoses*, Homer's epics, and Virgil's account of the aftermath of the fall of Troy, also reinforced a sense that virtuous behavior could be found in many people, not only among the right kind of Christians. La Mothe Le Vayer's *Vertu des payens* (*On the Virtue of the Pagans*, 1641) not only refused to condemn Socrates or Plato to the hellfires but expressed admiration for Confucius as well. And for those who were wary of worldly entanglements, neo-Stoic ethics were sweeping through Europe at the time, teaching that the chief goal of life, tranquility of mind (*apatheia*), could best be achieved by treating good and bad fortune as accidents to be overlooked rather than rewards or punishments to be grasped.[70] None of these moral views concerned themselves with the afterlife.

But Descartes himself would later condemn Stoic *apatheia* as well as the methods used by the Stoics to "conquer" the passions.[71] That would seem to place him among the followers of the materialistic philosophy of Epicurus, then flourishing thanks to the profound and wide influence of the great poem of his Roman follower Lucretius, *De rerum natura* (*On the Nature of Things*). Epicureanism taught that full acceptance of the world and its pleasures and pains provided a better model for moral behavior than renunciation.[72] Freethinkers sometimes also accepted that it was moral to go beyond dissimulation to pretense, saying what you really wished in private but in public saying and behaving according to who was listening.[73] While the opponents of Epicureanism argued that it taught nothing but an immoral pursuit of pleasure, its supporters found in its fundamental concern with the avoidance of pain a naturalistic foundation for ethical discipline and justice.[74] Even followers of Aristotle could become powerful advocates of materialism, as at the University of Padua, where the sixteenth-century philosopher

Pietro Pomponazzi promoted such views even though they could not offer reasons for the existence of an immortal soul, that problem being something for faith and religion rather than philosophy, he said. One of Pomponazzi's materialist successors, Cesare Cremonini, openly declared that the soul died with the life of the body. Balzac would write of "le grand Cremonin."[75]

Materialism implicitly undermined not only many tenets of Christian doctrine but also occultism and witchcraft. Giambatista della Porta wrote of "natural magic," but new philosophers like him explained the strange operations of nature as due to hidden or yet unknown causes rather than to the intervention of spirits or demons, much less the devil. Pomponazzi, for instance, condemned the prosecution of witches. Johann Weyer, physician to the Duke of Cleves, explained that even those people who freely confessed to witchcraft (the *lamiae*) were suffering from delusions, quoting Erasmus at length on "ecclesiastical gentleness" to advocate healing in such cases rather than punishment.[76] The nearby region of The Netherlands would be one of the first to suppress the prosecution of witches.[77] Descartes himself seems to have had relationships with the ruling family of the duchy of Lorraine, which was one of the hot spots for witch hunting, but the ducal family seems to have had nothing to do with the persecutions.[78]

In his own writings Descartes would retain a disciplined and meaningful silence on anything pertaining to belief in the active power of evil. He did famously evoke a "malicious demon of the utmost power and cunning" when describing how he might imagine that the things he knew about the world were merely delusions or dreams "devised to ensnare my judgement." But he makes it plain that this is a kind of literary device, a self-willed investigation of his own mind to explore the limits of doubt, not something that could be real. "In this human life we are often liable to make mistakes about particular things, and we must acknowledge the weakness of our nature." Yet "the exaggerated doubts" of his exercise "should be dismissed as laughable," he wrote in his *Meditations*. The purpose was to show that "no sane person" ever doubted that "there really is a world, and that human beings have bodies and so on," all derived from God's creation.[79] If God acted according to natural law alone, making the actions of arbitrary or evil spirits inadmissible, then demonic imaginings were due simply to inattention, error, or deliberate fiction. The latter possibility would not amuse many sincere believers, but eventually the refusal of learned judges to consider accusations of witchcraft in their courts would snuff out the imprisonments, brutal tortures, and grisly executions that marked the craze.[80]

The chief purveyor of ideas for the *esprits forts* (strong spirits), as they called themselves, was Giulio Cesare Vanini, who spoke for materialist Epicureanism in the France of Descartes's youth. Vanini had been a priest, was expelled from orders, and was received back into the Catholic Church before fleeing to Lyon and Paris to avoid the Inquisition. Finding refuge in the household of the grand François de Bassompierre—not only a libertine but a good friend of the young king and Marie de Medici—Vanini felt confident enough to publish his two major works, the *Amphitheatrum* (*Amphitheater*, 1615) and the even more radical *De admirandis naturae* (*The Marvels of Nature*, 1616). By quoting at length from many diverse ancient and Christian authors, ventriloquizing their words to advance his own views—he would, for instance, give in detail an argument of an atheist and then conclude by saying that he had refuted it elsewhere (when he had not)—Vanini raised doubts about the existence of divine providence. That led by implication to the rejection of miracles and the immortality of the soul, even the authority of scripture. Among his further conclusions were that "Nature is God," that human behavior is environmentally determined so that sin is a product of poor health, and that humans have emerged from nature and are no more than animals responding to the world via their senses.[81] To his freethinking friends he was a pitiless enemy of taboos, prejudice, and dogma. Recent studies of Vanini's life and work insist that he was neither an atheist, nor philosophically innovative, nor even a martyr to philosophy, but that he was a person of great learning with a refined and captivating way of speaking, and one of the more important voices in France of the school of Padua.[82]

But many considered Vanini and other intellectual libertines to be leading their sheep toward atheism. Even the great Erasmus could be condemned by the conservatives as "a falcon of atheism" for writing about the similarities among religions.[83] There were also rumors of a treatise circulating underground called "The Three Impostors," which argued that Moses, Jesus, and Muhammad—like some of the pagans before them—had tricked their followers into believing in their religious teachings as a way of seizing worldly power. When the anti-Aristotelian Dominican Tommaso Campanella was imprisoned by the Inquisition in 1595, one of the lines of questioning put to him by his torturers was whether he had authored that infamous work: he replied that it had been circulating for thirty years before he was born (putting it back to before 1540).[84] Other rumors associated the work's origins with the great twelfth-century Andalusian anti-Aristotelian philosopher known in Europe as Averroes (ibn Rušd); or his somewhat younger contemporary, the Holy Roman emperor Frederick II; or the fourteenth-century poet, Boccaccio; or the infamous early sixteenth-century Niccolò Machiavelli; or the printer, Étienne Dolet (burned at the stake in Paris in 1546); or the Aragonese Michael Servetius (burned at the stake in Geneva in 1553); or the ex-Capuchin monk, Bernardino Ochino (who died in Moravia in 1564 while fleeing persecution). We only have good evidence for the book's existence from the later seventeenth century, when the basic premise that religion was simply an empty but necessary political tool came to be associated with such "atheists" as Thomas Hobbes and Baruch Spinoza. But one of Descartes's mathematical friends, Claude Hardy, was reported to have held in his hands a printed copy of *De Tribus impostoribus* sometime earlier.[85]

At the time of Descartes's arrival in Paris, then, philosophical libertinism and purported atheism were sweeping through France, upsetting the authoritarians. It was not the kind of movement that agreed on a common set of teachings but instead took critique of received opinion as its most important strategy, coupled with methods of hiding one's true views because of the dangers of speaking otherwise.[86] Because one could speak honestly only with a few trusted friends behind doors, radical free thought often held common opinion in contempt, but in Paris a general libertinism in thought and behavior was common among those of Descartes's status.[87]

The mood in Paris would change. Following the seizure of the throne in 1617 by the devout Louis XIII—to which we will come in a moment—a period of powerful cultural reaction set in, with many of the leaders of the libertines being slandered, prohibited from speaking or publishing, or even threatened with their lives and liberty. Mersenne was among the many who wrote against them to bring France back to the true faith, later declaring that Paris was threatened by fifty thousand atheists.[88] Vanini himself was arrested in 1618 by the city government of Toulouse and charged with atheism, blasphemy, impieties, and other crimes, one of which was "bringing to light again the book intitled 'Of the Three Impostors.'"[89] His execution on February 9, 1619, proceeded by tying him to the stake, cutting out his tongue, strangling him with a cord, and burning his body so as to scatter his ashes to the winds (to prevent his resurrection at the end-time). Witnesses differed as to whether he died a screaming coward or a self-possessed gentleman, although they all remarked on the unusually strong bellowing he gave out when he refused to offer up his tongue voluntarily and the executioner seized it with tongs and sliced it from his jaw. In 1623, proceedings would also be launched against Balzac's friend Théophile for obscenity and blasphemy. He would be burned in effigy but later spend two years in prison before living his last under the protection of Henri II, duc de Montmorency, an associate of Gaston d'Orléans.[90] In a lesser but important action at the time, Descartes would defend Balzac himself from charges of immorality, acting as his champion.

But that was yet to come. In 1616, no one could see such a turn ahead. Balzac was certainly an Epicurean at the time.[91] And Balzac in turn had been close to Théophile, one of the intellectual leaders among the young libertines of the court, who published poetry that verged on the pornographic.[92] Descartes himself must have known all about the scandalous poet since not only had Balzac spent several years traveling with him but yet another of Descartes's intimate friends, the abbé Claude Picot, was close to one of Théophile's lovers, another poet (and former classmate of Descartes), Jacques Vallée, sieur des Barreaux. Among other matters, Théophile and Barreaux discussed the philosophy of Pietro La Sena, a true atheist.[93] Théophile's *Le Cabinet satyrique* (*The Cabinet of a Satyr*, 1618), reprinted five times, would later be pointed out as a prominent example of obscenity and blasphemy. A copy given to Queen Anne by Marie de Rohan was said to have led the former to become a flirt.[94] The one passage of poetry quoted by Descartes in his letters is from Théophile.[95]

For the moment, Descartes lived enjoyably among his friends, many of them libertine, probably visiting Gournay's salon, consulting with Mydorge, and seeking possible patrons among the rulers of the nation. He was keeping his options open. But in seeking a patron, Descartes was entering the rough and tumble of deadly earnest struggles for power. His hopes would soon be dashed through no fault of his own.

A Political Education

In Baillet's account, the narrative pauses following his comments about Descartes's early period in Paris, and he begins a new chapter. He recalls the political struggles of the day, which amounted to a period of continuing struggle for influence. The arrest of the Prince de Condé in September 1616 had been only the most dramatic. According to Baillet, Descartes had to be careful not to appear to be a partisan of the maréchal D'Ancre, who was opposed by a number of the greats.[96] Baillet further editorializes that serving among the opponents of the king would have been dishonorable. Descartes therefore thought of taking service with one of the king's close allies, either the duc de Guise or the comte d'Auvergne, but finally decided on leaving and joined the army of Prince Maurits in the Dutch Republic. What Baillet is gesturing at—somewhat misleadingly by referring to the king—is that the young Descartes was seeking to serve one of the nobles near the heart of the royal court, then led by Her Majesty the Queen Regent, Marie de Medici, ruling in the name of Louis. Marie identified with a former queen of France, Blanche of Castile, who had held the kingdom until her son came of age (he became known as Saint Louis): she had Blanche's portrait installed in her rooms and stated that her own son would be like that earlier holy king.[97] But her son was about to topple her regency in a bloody seizure of power. When he did, Descartes fled.

The identities of the people Baillet mentions place Descartes in or near the governing elite. The person from whom he needed to keep his distance, the maréchal D'Ancre, is better known as Concini, a nobleman of Florence who had come to France with Marie de Medici and risen to great state due to her favors. It was commonly supposed that Concini had secured her patronage by marrying the chief of her suite of ladies-in-waiting, her most intimate confidante since childhood, Leonora Dori (known as Galigaï). Concini had subsequently bought his way into the French nobility by purchasing the marquisate of Ancre, and in 1613 the queen had handed him the baton of a marshal of France (chief military commander), hence the titles by which Baillet names him. Concini had in effect become the regent's second in command, and possibly her de facto husband as well. Many of the great nobles despised him as an upstart who was ruining the kingdom for his own selfish ends. If Baillet says that Descartes had to be careful not to appear to be one of Concini's men, he may be suggesting that some people had suspicions that he was one. The implication of Baillet's report is that in seeking a position with the crown, Descartes had made himself known to some of the powerful, but given the circumstances, he needed to proceed with the utmost caution.

To avoid direct association with Concini, the young Descartes thought of serving with one of the other great nobles. One was the comte d'Auvergne: this would be Charles de Valois, better known by the title of duc d'Angoulême (a title he received later, in 1619). Charles was of royal blood, illegitimate son of King Charles IX and half-brother to Catherine de Balzac d'Entragues, Henri IV's most powerful mistress. (She had received a written promise of marriage from the king that was set aside for political reasons when Henri wed Marie de Medici in 1600.) Catherine had regained Henri's intimacy in 1604 and then plotted with her brother to obtain a declaration that one of her sons by Henri would succeed to the kingdom rather than Marie's eldest son Louis. For this conspiracy Charles was sentenced to death, but the sentence was commuted by Henri to life in prison. Charles had recently gained his liberty (in June 1616) by obtaining the favor of the queen regent. His and Descartes's paths would also cross from time to time during the next decade, especially in Germany but also at La Rochelle, where he initially commanded the royal army that the city fired on. Afterwards he continued to serve Richelieu loyally and led the French occupation of Lorraine. It would not be the relationship of a lifetime for Descartes, but in politics, even a few months can be transformative.

The other possibility was serving the duc de Guise, related to the Valois and whose interests seem to run through most of Descartes's early life. Baillet is strangely ambiguous here—perhaps deliberately—since the title might refer to either of two brothers, sons of the famous duke and great military commander who had headed the Catholic League during the religious wars in France in the later sixteenth century. The numerous members of the Guise family originated from the independent principality of Lorraine, at the time nominally subordinate to the Holy Roman emperor. But from the early sixteenth century the family had also come to serve the neighboring dynasty, the Valois, and had been awarded the French title of duc de Guise by François I. One of them, Marie, married into the royal house of Scotland, and by King James V had a daughter also named Mary; Mary Stuart was in turn raised at the court of France, becoming the wife of King François II and after his early death returning to Scotland to take up her position as Mary Queen of Scots, having a child of her own. Since the king of England descended from her line, the Guises were considered close cousins of both the rulers of France and of England. They were one of the most powerful families of Europe.

Despite the policies of his father, late in the sixteenth century Guise's eldest son, Charles, had helped secure the French throne for Henri IV in return for a huge sum of money and the governorship of Provence. He came to hold the position of grand master of France—head of the king's household—and he was a firm supporter of Henri's widow, Marie, the queen regent. His mother, Catherine of Cleves, was one of Marie de Medici's most trusted ladies and would later loyally follow her into exile at Blois. Charles also became very friendly with the king's younger brother, Gaston. But there is no evidence of Descartes's path deliberately crossing the duke's, while Charles had been an enemy of Balzac's patron, d'Epernon, in 1595. So it is unlikely that he is the person meant by Baillet.

More probably, then, Baillet meant to refer to Claude de Guise, also known as Claude de Lorraine, prince of Joinville, more commonly known by the title he was granted by the ten-year-old Louis in 1612, duc de Chevreuse. Chevreuse long continued as one of Louis XIII's favorites. He also later acted as Charles of England's proxy at his wedding to Henrietta Maria and headed the delegation that brought her to London, becoming a knight of the garter before his return. In 1622 he accepted the hand of Marie de Rohan in marriage; one of his younger sisters, the Princess de Conti, became another favorite of Queen Anne. Given Baillet's comment about how Descartes wished to join the king's troops, and Chevreuse's close relationship with Louis, he is the most likely candidate. In either case, Baillet associates Descartes with a lineage connected to Lorraine, and to Marie de Rohan, which helps explain some events in his future.

Perhaps a further hint lies in the later report by Descartes's friend and patron, Le Vasseur, that in the mid-1620s Descartes wore clothes of green taffeta.[98] Green was also the color in which Queen Anne of Austria dressed the men of her household, and the color worn by Gaston d'Orléans.[99] In the later 1620s, many friends of Descartes would be in the circle around Gaston—and in 1632 Gaston secretly married into the house of Lorraine over his brother's objections. In later years, after Descartes's death, the name of Anne, as queen regent of France (and mother of the future king Louis XIV) featured prominently on Descartes's tomb in Stockholm.[100] In 1666 she sent one of her trusted confidants to Stockholm to recover Descartes's body and bring it back for burial in his beloved Paris: that was done under the watchful eyes of Hugues de Terlon, a knight of Saint John. Anne also avidly collected holy relics, and although Terlon was allowed to keep Descartes's right index finger, he may have done so to pass it to her, since it does not figure in the register of his possessions at his death.[101] If, as is probable, Descartes later came to be a supporter of Anne of Austria or Gaston, that may well have been facilitated by the connection established in 1616–17 when he sought a relationship with one of the ducs de Guise, especially if it were Chevreuse, since his wife, Marie de Rohan, later actively plotted against Richelieu on behalf of her intimates Gaston and Anne. Such connections would explain Descartes's break with his father at the time of the execution of the comte de Chalais. The house of Lorraine had deep interests in the empire, too, where Descartes would be found in the early 1620s.

But we are getting ahead of ourselves. Whatever the details—which might one day be recovered—Baillet indicates that Descartes was seeking patronage among some of the great nobles who were close to Marie de Medici and the young Louis, but who were not close to Concini. We also know that Descartes's first period in Paris came to an end in the spring of 1617. As Baillet delicately puts it, an "accident" occurred at the end of April that dashed Descartes's hopes for further advancement: Marie de Medici's son seized power from his mother in a palace coup.

Together with his favorite and lover, Charles d'Albert, duc de Luynes, Louis decided to get Concini out of the way. On the morning of April 27, 1617, as Concini was making his way through a crowded room at the palace, the captain of the royal guards, the Baron de Vitry, informed him he was under arrest. When hands were laid on him, Concini shouted for help, at which point he was shot several times in the head and face and run through with swords; the mutilated body was stripped and hidden behind a staircase, and in the night it was secretly buried in a church under the organ. But a "mob" gained knowledge of where it was, broke into the church, dragged the body through Paris, and hung it by the feet on the Pont Neuf (which had been constructed by Henri IV), where they further desecrated the remains. Galigaï, Concini's wife and Marie de Medici's childhood friend, was also arrested, imprisoned, accused of sorcery, and beheaded and then burned in the Place de Grève. Marie de Medici's palace guards were replaced by Vitry's men. Distraught, and understandably feeling herself at risk, on May 3 the queen mother departed for the Château de Blois, where she would in effect continue under house arrest until fleeing with the help of D'Épernon. For the rest of her life, her relationship with her son would be fraught at best, despite several attempts to patch things up, with eventual death in impoverished exile. All Concini's rich estates in France and Italy ended up in the hands of the duc de Luynes. Concini's supporters were attacked by mobs or banished.

In this moment of deadly topsy-turvy, anyone who might be thought to have been a partisan of Concini's, or even someone who had aimed for patronage from the queen regent and her noble supporters, had to be extremely cautious. A new regime was muscling in. Descartes left Paris immediately, more or less at the same time as the former queen regent. The great nobles who had supported Marie de Medici were out of favor and running for cover, while there was enough of a suggestion about how Descartes had even been one of Concini's partisans for Baillet to have to deny it. Writing for readers who were living during the reign of Louis XIV, son of Louis XIII, Baillet would not have wished anyone to imagine that his subject had been caused any difficulties by the royal coup. As Baillet noted, that would imply disloyalty. He therefore avoided any discussion of the consequences of the coup for a young man who had been seeking to advance among those around the queen regent. Instead, Baillet insisted that Descartes had already decided to seek service with Maurits of Nassau in The Netherlands and had even collected his baggage in anticipation of departure before the "accident." The death of Concini simply did not change his resolution. Within a few days he was headed north.[102]

*

Dramatic events affect many people at once. Even a young would-be courtier such as Descartes, just turned twenty-one years old, was touched by the coup and would have to seek his fortune elsewhere. Whatever the details of Descartes's early connections and motivations, we can be sure that he had obtained an excellent education in the dangerous world of the French court. He was good at cards, but the hand he had been dealt would make for no easy game. It would cause him to remain on the move for the rest of his life, mainly abroad.

PART 3

Gearing Up for War

Mathematical Inspirations

His fortunes for the moment blocked in France, Descartes decided to learn the art of war, which was about more than the personal ability to fight. The nobles of the sword who ruled his country all had experience of military campaigning, as did most of their courtiers, and campaigns required engineers. The younger René could have joined one of the factions fighting in France, but that would have committed him to a party at a moment when the future there was quite uncertain. His new friend Balzac had visited the Dutch Republic in recent years, and had much to say about it; it was then in a period of truce with Spain but had been fighting for decades to obtain and retain its independence from Madrid, so it was in the low countries where the latest methods of war had been developed to a high point. The French king Henri IV had concluded an alliance with the Dutch Republic (and England) in 1596, and nearby Lorraine had long-standing interests there, too. The sieur du Perron would therefore keep his options open and head north. Events and opportunity would further conspire to maneuver him into the opening stages of the Thirty Years' War, serving in part as a military engineer. Almost by accident, then, he also became an accomplished mathematical practitioner. But he never abandoned his love for the true and imaginative spirit of poets.

Breda

Descartes set out to study the art of war. The seventeenth-century biographer Adrien Baillet's words are *apprendre le métier de la guerre*, and *métier* has an implication of a skill that employs knowledge.[1] The word might have applied to learning the bodily skills of military drill, which were certainly complex, but Baillet more probably meant that Descartes intended to learn about the engineering-related skills necessary to large-scale armed conflict, for he immediately adds that the young man went to study under Maurits of Nassau, and he was renowned as one of the most successful generals of his generation because of his use of the latest technical methods in his campaigns.[2] In going to the Dutch Republic, Descartes would have been learning his *métier* with some of the best practical military mathematicians in Europe.

He is known to have been at the main garrison of Breda, but perhaps he did not go there straightaway. Despite what Baillet wrote about how Descartes had determined on a plan to serve Maurits of Nassau, the elected commander in chief (*stadholder*) of the Dutch Republic—how Descartes had even packed his bags before Concino Concini's assassination in April 1617—antiquarian researchers later discovered two baptismal certificates signed by Descartes at the end of the year, in Sucé, near Nantes, where his stepmother had a house.[3] There is also a recorded comment Descartes made many years later to his Dutch friend Frans van Schooten, saying that he had lived in Breda for fifteenth months; if one knows the date he left Breda (which is contained in a letter he wrote to his friend Isaac Beeckman on April 29, 1619), that puts his arrival in the city as shortly after the New Year in 1618.[4] Coincidentally, in February 1618, Maurits finally inherited the title of Prince of Orange and Baron of Breda on the death of his brother, Philip William. The careful and knowledgeable Baillet says that Descartes wished to serve "le prince."[5]

It is therefore likely that following the coup in Paris, Descartes went to visit his father and family before making a decision about his next move. In accordance with French custom, Descartes would reach the age of his majority in March 1618; in July his older brother Pierre would obtain legal permission, with his father, to take over the estates that came down through their mother's inheritance, including the seigneury of Perron, probably settling an income on René.[6] René probably needed to be assured in advance of means, and perhaps to arrange for transfers of money between someone along the French Atlantic seaboard and the Dutch Republic. Only then did he set out north.

It is possible, too, that Joachim Descartes did his son the service of introducing him to Philippe de Béthune, comte de Selles. Béthune happened to be the younger brother of the famous *politique*, the duc de Sully, King Henri IV's minister and the recorder of the story told in part 2 about the Spanish spy and the Descartes who was probably René's father. Philippe de Béthune had converted to Catholicism but remained among the *politiques*, and he had represented France as ambassador to Rome from 1601 to 1605. He also happened to be the royal administrator (*lieutenant général*) of Brittany and governor of Rennes.[7] We can therefore assume that from his parliamentary activities, at least, Joachim Descartes knew Béthune; René's path would cross Béthune's several times in years to come; perhaps the father had presented his cultivated and ambitious son, recently in Paris, to the count.

The next sign of René is from the Dutch Republic. It comes many months later, in an entry in the notebook of Isaac Beeckman on November 10, 1618. Beeckman himself had arrived in Breda just three weeks before then, after having taken his medical doctorate in Caen, in Brittany, but he would become an important figure in Descartes's life. The entry records how a Frenchman from Poitou showed Beeckman an attempt to prove that a certain angle did not exist. Later entries make it clear that the person in question was the sieur du Perron, René Descartes.[8] When they had first met we do not know, nor do we know how long beforehand Descartes had been resident in Breda. But November 10 is a firm date.

Most of his biographers agree that Descartes probably went straight from France to the chief military settlement of the army of the Dutch States General at Breda, where two French regiments were based because of the alliance. Years later, according to another early biography, Descartes would sometimes show his friends a souvenir coin he kept, which he had taken in payment for military service. (Following Roman tradition, such a coin was termed a *solde,* which gives us "soldier.") But the coin he kept was reportedly a Spanish gold doubloon, more likely to have been used by the imperial troops with whom he later served than by the States Army of the Dutch Republic, which paid in silver.[9] Most likely, then, Descartes was in Breda as a gentleman volunteer, as was common for his social rank, serving at his own expense and acting as he thought best under the leadership of the general in charge rather than taking orders from anyone else. Early in the twentieth century Gustave Cohen investigated the archival sources related to the French regiments, and he did not find Descartes on their rolls. Descartes himself would write a note to his mentor, Isaac Beeckman, describing himself as "without occupation or office," which certainly implies that he had not enlisted.[10]

Perhaps, then, he had not formally joined the French regiments after all. Baillet himself states that Descartes served with Prince Maurits. If so, that would have given him a chance to learn about armies on the move rather than in camp despite the ongoing truce with the Spaniards. Descartes said something about it many years later, when he was sharply criticized by Dutch opponents in Leiden and pushed back by claiming that he had carried arms to help the Dutch keep out the Spanish Inquisition.[11] The French also certainly objected to the Inquisition: its remit did not run on their territories. But if Descartes "carried arms" in this way, he could only have been saying that he had helped the prince's forces maintain order in the Dutch Republic at the time, reminding his opponents of the common Dutch opinion that the Spaniards would have taken over through subversion had it not been for the troops of the prince.

In fact, when Descartes arrived in the United Provinces, the country was torn by a conflict that verged on civil war. The political struggle pitted the urban magistrates who thought of their polity as a patrician republic led by themselves against those who thought that only a strong prince could unite the factions within their state and advance their cause abroad. Such conflicting opinions had come to a boil in 1617–18 because of a theological dispute over free will. Attempts were made to require all ministers to publicly support the stern declaration of Calvinist faith, the *Confessio Belgica*, which declared that personal salvation is predestined by the grace of God rather than by freely performed good works. But a powerful group objected to the political attempt to impose such a theology on all, both because of concerns of compelling consciences and because many of them held a less-deterministic view about salvation. These Remonstrants (so called because they had "remonstrated" against the ordinance) were supported by the most powerful member of the States General of the United Provinces, Johan van Oldenbarnevelt, who was backed by the legal-political theorist Hugo Grotius, sometimes considered the founder of international law. On the other side were the Counter-Remonstrants, a group eager for clear and precise theological definitions to determine office, so that anyone with views that might be considered sympathetic to Catholic opinion about free will would be excluded from power. The Counter-Remonstrants were supported by many of the nobility along with the urban militias and guildsmen who thought of Rome as working tirelessly with Spain to undermine their liberties by secret plots.

The disputes had been growing since 1610, but at the end of 1616 and beginning of 1617, the "mobs" often allied themselves with the Counter-Remonstrants against the urban patricians. Economic stresses had caused generalized social unrest to break out in many places, but the theological divisions gave them an ideological edge. To protect their own interests, then, the magistrates authorized the raising of special municipal troops (the *waardgelders*), who swore allegiance to their provincial governments instead of to the national States General.

On the principle that there could be only one chain of command, Prince Maurits at last openly sided with the "patriot" Counter-Remonstrants. Over the course of the summer of 1618, he openly moved against Oldenbarnevelt's party, marching the States' army from city to city and forcing the *waardgelders* to lay down their arms, which also allowed Maurits to purge the urban governments of Remonstrants in favor of his own loyalists. Oldenbarnevelt and Grotius were arrested in August. In the spring of 1619, the distinguished Oldenbarnevelt was found guilty of treason; his head was cut off by the executioner soon thereafter. (Grotius was smuggled out of prison by his wife and would live out the rest of his life in exile.) In the meantime, Maurits allowed a national synod to be called to decide on the theological issues, which met in the capital of the province of Holland: it sat from November until May in Dordrecht (or "Dort"), the famous Synod of Dort that finally also declared in favor of the Counter-Remonstrants.

Descartes's known period of service in the Dutch Republic therefore coincided with the period when Maurits was seizing power in alliance with the Counter-Remonstrants. Baillet later reviewed the nature of events at the time and remarked that Descartes could have served Maurits on his brief campaign but had instead remained in the Breda garrison.[12] But he did not know of Descartes's later comment about helping to fight off the Spanish Inquisition. Descartes could have been speaking to his critics disingenuously, but if he had ever been caught in a lie, it would have made things even worse for him. He was probably, then, reminding them of how he had taken up arms in their cause. That would mean that he had served among the troops who forcibly disbanded Oldenbarnevelt's *waardgelders*. As a matter of fact, he apparently took their part, too. Descartes later told a student of philosophy that "after weighing the truth of the matter, the author finds himself in agreement with the [Counter-Remonstrants], rather than the [Remonstrants]," even though it was against what he had been taught by the Jesuits.[13] Perhaps he was merely speaking in the language of the student, but it indicates sympathy for the position of Maurits. When Descartes left The Netherlands, it would be on the verge of the conclusion of the Synod of Dort in favor of predestination.

Military Engineering

But serving the prince at that moment does not mean that Descartes spent all of his time with hand on sword. He had gained experience with an army on the move, but by the latter part of 1618, with the Prince of Orange now secure and winter coming on, the French volunteer could seize the opportunities to comprehend the *métier* of war that were on offer in the heavily fortified country. That meant learning to handle the problems and instruments of practical mathematics, which flourished in The Netherlands (see fig. 7).

By the time Descartes went north, the soldiers of the low countries had had more than four decades of almost continuous experience with war. It helped make them fine mathematicians. In France, around 1600, King Henri IV considered founding schools that would train the sons of the nobility in mathematics and fortification, probably prompting him to establish the *collège* (school) at La Flèche, among others. In Descartes's own generation, however, the military instruction of the *collège* had not yet become strong; indeed, there is no evidence that he had much instruction in mathematics at all until after he left school.[14] Some nobles—including the duc de Luynes—took a special interest in encouraging military engineers with their patronage. Descartes himself became well acquainted over the years with the three most prominent French military engineers of his time: the comte de Pagan, Pierre Petit, and Girard Desargues.[15] In the future, the French state would certainly prosper in part due to its engineers, one of whom (Napoleon) became emperor of all Europe.[16] Elsewhere in Europe, too, men with the practical scientific and engineering abilities necessary for gunpowder warfare could rise to great heights.[17] But at the time, for an intending student of war, the best education was to be had in the low countries. There, a long and desperately fierce conflict—it would be called the Eighty Years' War—had been taking place in a region so heavily urbanized that there was little room for a war of maneuver, which meant that war was waged at sea and overseas, and by siege. By necessity, the people of the low countries had learned how to excel in the business.[18] The low countries

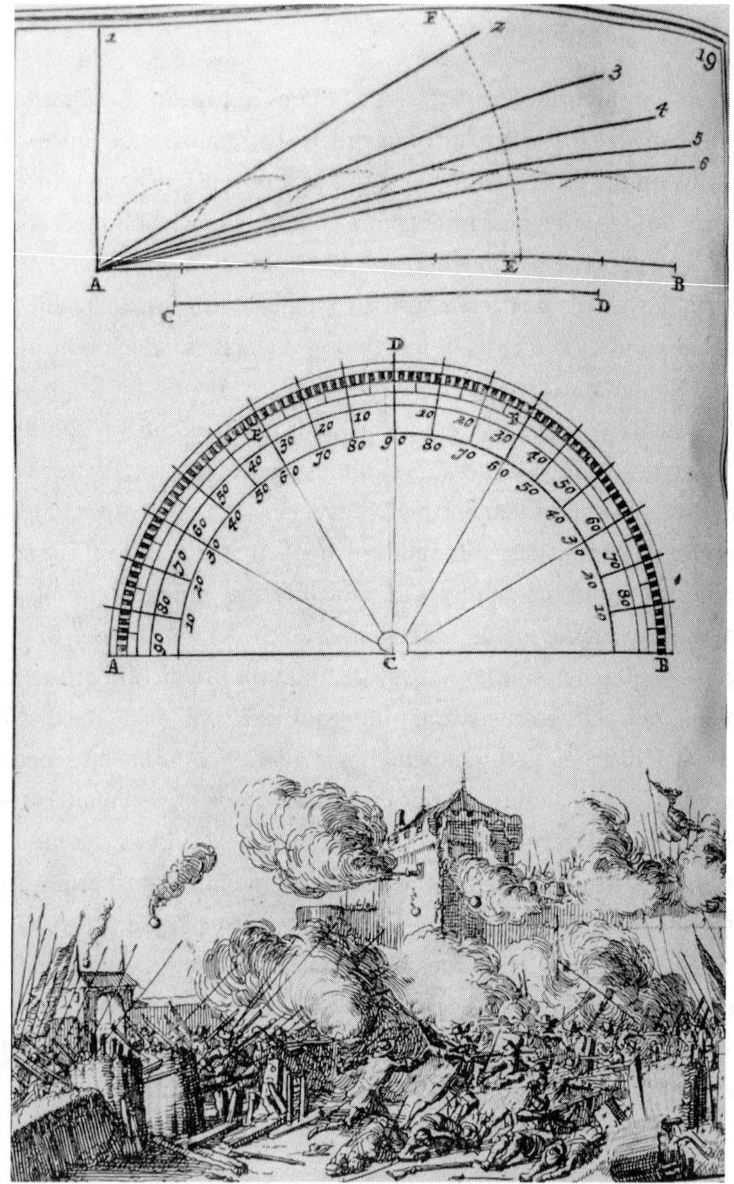

Figure 7. Geometrical figures and assault on a fortress. Reprinted from *Den Arbeid van Mars* (Amsterdam: Jacob van Meurs, 1672). Courtesy of Anne S. K. Brown Military Collection, Hay Library, Brown University.

became the international nursery of war, where such people as the famous French commander Turenne would learn his art.

War making was not simply about organized violence. By the beginning of the seventeenth century it had become a complex applied science.[19] Its study focused on defensive bastions and offensive siege works, the lifting of great weights and the strength of materials, accurate surveying and measuring of heights and distances from afar, waterworks and mining, and in general the application of the latest technological devices and mathematical methods related to complex machines (from pulleys to mills) to solve physical problems. Such moves had become necessary from the later 1400s, when new methods for casting iron and bronze guns made it possible to produce canons of up to 35 tons capable of firing two-hundred-pound balls long distances. If such weapons could be hauled into place, they could smash any medieval wall, as the Turks did at Constantinople. Refinements in the making of gunpowder soon made it possible to wreck the old curtain walls with smaller-bore guns, too. With the new technology of mobile artillery, the French swept through the castellated cities of Italy in 1494–98.[20] The architectural response was the "trace italienne," a strategy for designing defensive walls that could withstand bombardment. The walls were relatively low, wide, sloped, and stone-faced, laid out in a star-shaped pattern to offer attacking guns only angled surfaces that would deflect fired missiles. Backed by great earth-filled ramparts onto which defensive canons could in turn be lifted, and ditches and dikes out front, the new defensive structures required attackers to approach with angled trenches and complicated mechanical siege engines, or to mine underneath the walls so as to set off an explosion that would create a gap through which soldiers could pour.[21] In the 1570s and 1580s, Adriaen Anthonisz of Alkmaar, known as Metius, constructed new-style defenses (fig. 8) for a number of northern Dutch cities that added waterworks to the bastions, developing what became known as the Old-Dutch system of fortification.

The resources required for mounting both offensive and defensive

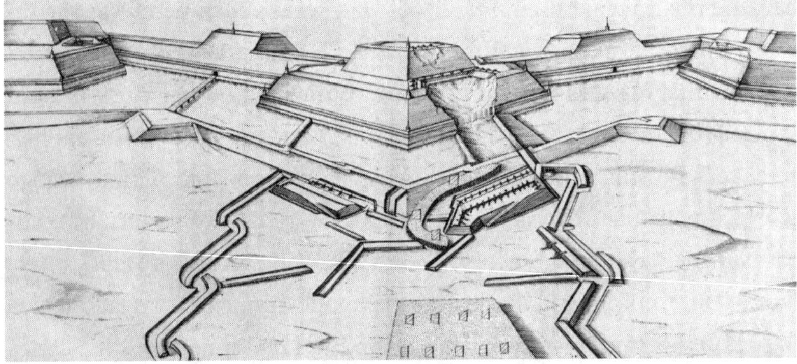

Figure 8. Idealized assault on new-style fortifications (note battery, center right, supported by entrenchments, demolishing a bastion). Reprinted from Hans van Schille, *Form vnd Weis zu Bauwen, Zimmern, Machen* (Antwerp: Gerardum de Jode, 1580). Courtesy of Anne S. K. Brown Military Collection, Hay Library, Brown University.

war became astronomical. But the methods of both attack and defense required exact engineering and quickly calculated responses by competent experts. The officers in turn needed to know not only how to work with weights and machines, as well as technical drawing, perspective, and the appropriate methods and devices to measure volumes (stereometry), but also how to use the mathematical instruments required for surveying, cartography, and gunnery. From before Leonardo da Vinci's time, humanist writers had argued that learning how to draw in three-dimensional perspective (*disegno*) was not only important for the cultivation of the mind but also "very necessary, as much in this profession of fortification as in all the other arts."[22] For naval warfare, which used wind to float batteries of artillery from place to place on the water, navigation was a further requirement.[23] Mathematics was even transforming swordplay, with geometrical methods of fencing coming to the fore, and we know that at the end of his life Descartes left behind a treatise on fencing that has since gone missing.[24]

The person who advised Prince Maurits most closely on many of his engineering projects was the famous Simon Stevin, perhaps now best known for popularizing the use of decimals as a substitute for fractions. By the time of Descartes's arrival, Stevin was about seventy years old, but he remained nominally in charge of the engineers of the French regiments serving the Dutch Republic.[25] Stevin had become closely associated with Maurits around 1590 after already making a reputation for designing all kinds of ingenious devices; he then became Maurits's tutor on a wide range of practical subjects and on the uses of mathematics in their solution. For instance, Stevin taught Maurits how to use methods of double-entry bookkeeping for regularizing governmental accounts, arguably one of the most important innovations of the era. Most of the many works he printed were first composed for the prince, including his *Sterctenbouwing* (*Stronghold Construction*) of 1594, which codified what became known throughout Europe as the Dutch Fortification System.[26] In 1600 Stevin and Maurits also turned a fencing academy at Leiden into the Duytsche Mathematique—the engineering school.[27] It turned out several generations of practical engineers and political leaders who continued to use and promote mathematics. They helped develop, for instance, probability theory. And when campaigns were under way, Stevin was constantly at his prince's side.[28]

In fact, Maurits has often been credited with bringing a military revolution into being.[29] He introduced new kinds of military formations derived from Roman examples, which required a great deal of discipline and prior drill on the part of the soldiers: indeed, it may be that the most important innovation was causing the various kinds of soldiers to perform their particular tasks in coordinated small units rather than acting en masse.[30] Maurits also mixed musket with pike, armed his cavalry with wheel-lock pistols, made sure to have his soldiers paid regularly so as not to cause them to prey on civilian populations, and furthered many other innovations. He was not only an able commander but also an enthusiast for applying mathematics to solving all kinds of problems, apparently saying on his deathbed, when asked about his faith, that he believed that two and two were four. He knew as well as anyone that only careful preparation and planning, and proper materiel, permitted successful actions of valor.

Descartes's encounter with engineering would have a profound effect on his thinking.[31] He even begins a well-known passage in his *Discours* (1637) with a military example. In it, he explained why it is best to build one's thoughts afresh rather than to try to work with old ones. He was on his way back to the army at the beginning of winter, he writes, and

> Among the first thoughts occurring to me was that there is not usually so much perfection in *works* composed of several parts and produced by various different craftsmen as in the efforts of one man. Thus we see that *bastions* undertaken and completed by a single architect are usually more attractive and better planned than those which several have tried to patch up by adapting old walls built for different purposes.[32]

The passage is even stronger if one remembers that in his day "works" often meant a fortification, or a part of it, as in "earthworks" or a "hornwork." "Bastions" (*bastimens*) refers to defensive fortifications, where any simple abutting of different constructions would be a place of weakness certain to be pummeled by the enemy's guns. Moreover, he goes on, "orderly towns" based on geometrical designs were being built anew in his day by "engineers." (He must have been thinking of newly built cities in colonial territories or such new European cities as Glückstadt, or possibly even the new urban residential squares such as the Place Royale in Paris.[33]) But even in older cities there "have always been certain *officiers*" whose job it is to make sure "that private buildings embellish public places." From examples like these, "you will understand how difficult it is to make something perfect by working only on what others have produced."[34] From such musings he drew the lesson that he should begin his own philosophy from his own first principles. Descartes's metaphor about starting afresh begins with a moment of reflection after time serving with an army, thinking about fortifications.

Meeting Isaac Beeckman

Assuming that Descartes served Prince Maurits during the campaigning season of 1618—perhaps he even caught sight of Stevin himself—he must have returned to Breda after the summer action at the latest, for that was where he met Isaac Beeckman in early November. In the view of most commentators, this was the most important intellectual turning point in Descartes's life, for his new friend was not only an able mathematician adept at applying mathematics to solving physical problems but also an experimentalist, physician, and philosophical atomist. (Like Descartes, he also supported the Counter-Remonstrants.[35]) Together they worked out a number of problems, including the law of motion that now goes by the name of inertia. Descartes himself later wrote to Beeckman that "truly, you alone have roused me from my idleness and recalled to me what I had learned and already almost forgotten. When my mind had strayed so far from serious occupations, you led it back to better things. Therefore, if by chance I produce something of merit, you can rightfully claim it as yours."[36]

An account of their first meeting was reported by Daniel Lipstorp, a native of Lübeck who probably had it from two of Descartes's Dutch friends, Johannes de Raey and Frans van Schooten.[37] Lipstorp wrote that while in Breda, Descartes noticed that a mathematician had posted a difficult problem in a public spot with an invitation to anyone to propose its solution. There were many private mathematical tutors in the period, and they often tried to attract pupils by advertising their superior abilities in this way, showing that they could solve difficulties no one else could.[38] Descartes stopped to have a look, but "because he had arrived in the Netherlands only a short time before," he could not understand the sheet well (the words being in Dutch) and asked for help in French or Latin. It was Beeckman who responded from the crowd. Descartes must have indicated that he knew the answer, for Beeckman in turn asked Descartes to write down his solution and bring it to him. Descartes later did so—just as Viète had done some years before, Lipstorp adds—and earned Beeckman's respect. So began the relationship.

Several things are suggestive about this story. First is the comparison to François Viète, one of the most famous French mathematicians of the period: some consider him to be the inventor of symbolic algebra because of what he published in his *Isagoge* of 1591 for his pupil

Catherine de Parthenay.[39] The episode to which Lipstorp alludes is this: a well-known Dutch mathematician, Adriaan van Roomen, had circulated a difficult problem that no one else could solve; when the ambassador from the Dutch States General was visiting the French king Henri IV, who was fond of expounding on the merits of his people, the Dutchman noted that none of the king's subjects had yet been able to answer his own Van Roomen; Henri then asked Viète, who was with him at Fontainebleau, to have a look, and he came and immediately gave an answer. Viète soon became even more famous for solving an ancient difficulty raised by Apollonius of Perga.

Given this well-known story, the comparison between Viète and the young Descartes has made some commentators consider it to have been a model for the meeting rather than a report of what happened. Van Schooten, who may have told the story to Lipstorp, later edited the works of Viète as well as Descartes, so the episode might hint that Van Schooten thought that Descartes was later walking closely in Viète's footsteps without acknowledging his predecessor.[40] But he might also be drawing our attention to their common backgrounds: both were from Poitou, both came from families of lawyers and merchants, both were known at the time by the titles of their siegneuries (in Viète's case, seigneur de la Bigotière), and Viète, like Descartes's father, served as a *conseiller* in the *parlement* of Brittany in Rennes: perhaps the reputation of his father's colleague was even an inspiration to the young Descartes. Viète spent much of his life living among, and serving, the monarchy and great nobles of France, including the Rohans.[41] Like Descartes's father, Viète even had experience with espionage, in his case decrypting a complex and "infallible" Spanish code in 1590.[42] The parallels are certainly worthy of remark.

But second, the story of the origin of the meeting of Descartes and Beeckman suggests that Descartes was looking for a mathematical tutor and that Beeckman was willing to help. Beeckman seems to have had a bit of time on his hands: he was in Breda to assist his uncle during the slaughtering season, as well as to court a young woman.[43] He left Breda at the end of 1618, only a few weeks after meeting Descartes, traveling thereafter for some time, presumably looking for a suitable place to set up practice as a physician.[44] (It would take Beeckman another year before he found regular employment, as second to a friend in running the Latin school in Utrecht.[45]) In the meantime, he might well have been willing to supplement his income by taking on pupils for tuition, just as he himself had first learned mathematics in a private three-month course.[46] The fact that in Lipstorp's account Beeckman steps forward from a crowd near a public notice about a mathematical problem suggests that he may have been keeping an eye out for clients; even more suggestive is his reported invitation to Descartes to write up his proof and bring it along to his address—which he gave him. It sounds as though he was at least open to taking on pupils. Soon after their meeting, Beeckman asked Descartes to compose his thoughts on music "for me" (*mea gratia*), indicating that he thought of himself as Descartes's mentor rather than simply his friend.[47] That Descartes later promised Beeckman to put his *Mechanics* or *Geometry* in order also implies a promise to sort out his notes from their time together.[48]

Beeckman certainly had much to offer. He had prior experience with solving practical and technical problems and had worked in mathematics and music as well, but his most advanced recent studies were concerned with the relationships between nature and the human body. Although we now often refer to this field of study as "medicine," at the time it was termed *physique* in French (*physic* in English), from the Greek word for nature, *phusis*; only in the nineteenth century would the word *physics* come into common currency to indicate a distinction between the study of nature and the study of bodies. Like others, then, Beeckman "chose medicine as a way of studying nature systematically."[49] It was probably from his medical studies that he became a confirmed supporter of the view that all natural things are composed of tiny physical particles.[50]

The other major topic on which Beeckman was writing in his journal at the time was related to the last corollary in his thesis—music—which would be the subject of Descartes's first formal piece of writing.[51] Since antiquity, music had been considered one of the mathematical sciences, but in Beeckman's generation it was going through a revolution of new experimentation and theorizing about nonlinear proportions and dissonance, in which the Venetians Gioseffe Zarlino and his pupil, Vincenzo Galilei (father of Galileo Galilei), played an especially important part.[52] These moves would lead to new manners of musical expression and form, as in that of Claudio Monteverdi. Beeckman had been working on such problems himself. The treatise on music that Descartes sent to Beeckman on the last day of 1618—and which was published posthumously—shows the pupil to be knowledgeable about overtones and sympathetic vibration, for instance, which were quite cutting edge at the time, and it no doubt therefore owed a great deal to his mentor.[53] But Beeckman was seeking both a place in the world and a wife. Although Descartes sent him the musical treatise on December 31, his teacher had already left Breda. Beeckman would receive his copy on January 2, 1619, on his way to his homeland of Zeeland.[54]

A further indication of Descartes's new interests is a manuscript he was preparing on hydrostatics that he mentioned in a letter to Beeckman. It tried to give an explanation for a formal proof that had been stated by Stevin, and it involved the mechanics of the corpuscles that made up the fluid. It was probably inspired by Descartes's conversations with Beeckman but would also have been in keeping with the epicurean atomism he had already encountered in Paris.[55] According to the fragments of Descartes's writing that Baillet places in this period are a group under the heading of "Democritica" (no doubt referring to the ancient moral philosopher associated with Pythagoras), and a work on *The Mind of Animals* demonstrating how their knowledge is entirely a product of their physiology.[56]

In other words, we may suppose that with the army now in winter quarters, Descartes took up the study of the art of war as practiced in the low countries and looked for help. At the time, soldiers and officers were quartered in the homes of townspeople (who charged the government rent for the service), and they would have had much time on their hands, which Descartes put to good use. For a few weeks he had the help of an inspiring person, but Beeckman was quickly gone. At the end of the little musical treatise Descartes had written for him, Descartes rather plaintively noted that it had been composed "in the midst of turmoil and uneducated soldiers, by a man . . . busy with entirely different thoughts and activities."[57] He soon immersed himself further in practical studies. On January 24, 1619, he wrote to Beeckman that he was studying *pictura* (meaning *disegno*, or perspective drawing, important for mathematical practitioners as well as artists), military architecture, and "especially" the Dutch language. He apologized that these were not the kind of "elevated subjects" with which Beeckman usually occupied his mind, but they were useful.[58] The subject matter suggests that Descartes was busily at work studying the kind of topics taught in the engineering school at Leiden, which were delivered in the Dutch language (as were the textbooks produced to go with them), and which included perspective drawing.

But success with the new mathematical engineering was not simply an intellectual problem. Years later Descartes would comment that mathematical knowledge "is not to be gleaned from books, but rather from actual practice and skill. The author [i.e., Descartes] had to teach himself the subject that way, since he had no books with him, and the results he obtained were very happy."[59] Learning by practice and skill meant learning how to use instruments. Military engineers of the era were not only changing the face of battle but also developing the modern ability to construct physical worlds by human design using mathematical tools.[60] A famous example is Descartes's older contemporary, Galileo, who had begun by following in the footsteps of such artisan-engineers as Leonardo and Niccolò Tartaglia (perhaps best known now as the modern editor of Euclid).[61] After obtaining the chair in mathematics at the University of Padua (in 1592), Galileo opened a workshop in his house to teach the use of machines within fortresses and to construct (and sell) mathematical instruments that would allow officers to

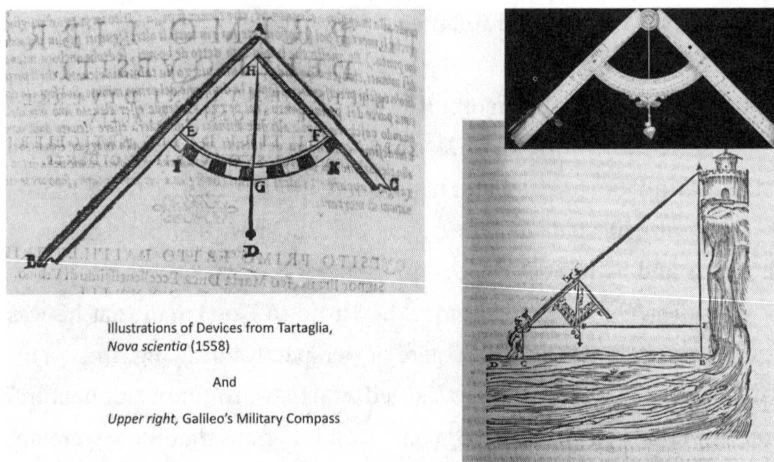

Illustrations of Devices from Tartaglia,
Nova scientia (1558)
And
Upper right, Galileo's Military Compass

Figure 9. Mathematical Instruments for military engineers, including Galileo's military compass. Reprinted from: Niccolò Tartaglia, *Nova Scientia* (In Vinegia : [s.n.], 1558), courtesy of Lownes Science Collection, Hay Library, Brown University; "Galileo's Compass—History of an Invention," Museo Galileo, Florence, http://brunelleschi.imss.fi.it/esplora/compasso/dswmedia/storia/estoria1_st.html.

perform quick and accurate calculations, particularly his "military compass" (fig. 9).[62] Working with instruments and the perspective drawing technique of *perspectiva* also encouraged his interest in optics.[63] In 1610 Galileo would become more famous for turning a novel military instrument for seeing at a distance—the spyglass, or telescope—onto the heavens, earning him an international reputation as an astronomer and natural philosopher and giving him the opportunity to become philosopher to the duke in Florence.[64] Instrumentation would be critical to the new science.[65] The resulting new forms of mechanical calculation even affected the development of landscape painting.[66]

In Descartes's case, the result of his studies with instruments would be a further revelation. In a letter of March 26, Descartes wrote excitedly to Beeckman about how, "with the aid of my compass," he had found "four remarkable and completely new demonstrations."[67] The compass here would have been something similar to Galileo's military compass, a version of a device known as a sector. It consisted of two straight arms that were ruled and inscribed and joined at one end, allowing the arms to be separated at the other end, usually held together by a curved piece of metal that was also carefully marked to indicate angles. Various rules could be placed on each arm, and by comparing them, complex calculations could be performed relatively easily. The sector was something like a twentieth-century slide rule and drawing compass combined. It could also be used for finding heights, depths, and distances, the size of shot, the angle of trajectory, the number of men who could stand within a space on the ground, and even the conversion of one monetary currency into another (very useful on foreign campaigns). But Descartes was working with one of the "new compasses" that had the additional ability of drawing continuous but changing curves. Perhaps it was similar to the instruments that Descartes's later friend Frans van Schooten illustrated, which could be used to draw ellipses (Van Schooten later edited Descartes's *La Dioptrique* and may have been the source for the story of the meeting with Beeckman).[68] Van Schooten's father also happened to be a professor in the Duytche Mathematique, the Leiden engineering school.

Using one of the new compasses, then, Descartes had found a method for dividing an angle into any number of equal parts, as well as how to work with three kinds of cubic equations. He had hit on a method for moving beyond geometry alone and even thought that he was on the verge of finding a new way to obtain roots. He was, in other words, seeking "a completely new science" that united both continuous and discrete quantities, geometry and algebra, in the service of solving problems.[69] Put another way, he was finding new methods for a unified understanding of how shapes and equations can be produced from one other: his new understanding involved defining curves "in terms of the motions that generated them,"[70] as in something we still call "Cartesian coordinates." "Through the confusing darkness of this science I have caught a glimpse of some sort of light," he wrote to Beeckman. He even imagined for a moment that he might be able to construct a new kind of large but exacting instrument that could find variations in the

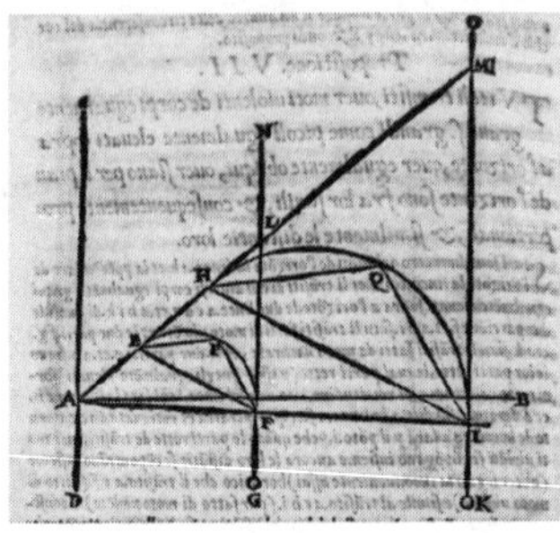

Tartaglia, *Nova Scientia* (1558)

Descartes, *Geometrie* (1637)

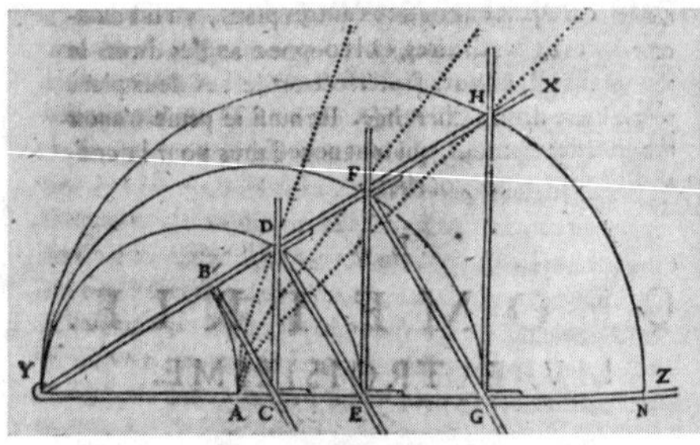

Figure 10. Geometrical curves, calculated; geometrical curves, drawn with an instrument. Reprinted from: Niccolò Tartaglia, *Nova Scientia* (In Vinegia : [s.n.], 1558); René Descartes, *Geometrie* (Leiden: Ian Maire, 1637). Courtesy of Lownes Science Collection, Hay Library, Brown University.

moon's course through the stars, from which the problem of calculating longitude could be solved, although he soon came to realize that the device would be impractical. In a letter of a month later he confirmed that "I really did discover" these new methods "with the aid of the new compasses."[71] The mathematical instruments of war were changing Descartes's mental landscape (see fig. 10).

At the new year, Beeckman presented his pupil with a gift, a notebook in which Descartes began to record his thoughts, as his teacher had done in his "journal" some years earlier.[72] They clearly had become close. After Beeckman left Breda, Descartes continued to write to him, clearly wanting to share his thoughts. In the letters, Descartes says that Beeckman had introduced him to a new way of thinking that Descartes continued to develop. They also express his love. For instance, he wrote that he would put his new language skills to work by making a trip to see Beeckman in Middleburg at the beginning of Lent, signing off with "an unbreakable bond of affection."[73] He made the visit on about March 20, but he was unable to find Beeckman.[74] Consequently, the young nobleman sent his mentor the letter that excitedly explained what he had found in the past few days with the compasses. The letter commands "Love me," and speaks of "bonds of everlasting love." Since there are no examples of other letters of Descartes in the period, it is hard to judge whether his language indicates some sort of formal but intimate courtliness, or the relationship of an orphan to a newfound father, or homoerotic feelings, any or all perhaps provoked by the fear of having lost his interlocutor.[75] Beeckman did respond, although in a friendly but more distant manner. In retrospect, however, thanks both to Descartes's love for his mentor and to the new instrumental mathematics he studied privately, a new world had opened up to him.

The Holy Roman Empire

In the spring of 1619, however, electrifying news had spread from farther east, and armed men were listening. Between his letter of March and the next one, on April 23, Descartes had heard. "Do not expect anything from my Muse at the moment," he wrote Beeckman, "for while I am preparing for the journey about to begin tomorrow, my mind has already set out on the voyage." He had not yet been "summoned" to Germany, but he heard that armies were being enlisted, even if there would not necessarily be any fighting. It was his moment to be at the center of things. Destiny called, and he was preoccupied with anticipation. He was also assembling useful information: if events continued to move in the direction he expected, he would head north to the Baltic and make his way east to Poland, then south to Hungary, since that would be the route most likely to take him to the scene of the action without encountering "marauding soldiers." With spring in the air, new images preoccupied his mind, ones that were fully personal, and grand.[76]

The news had not been unexpected. The emperor Matthias had long been ill, and he finally expired on March 20, 1619—the same day that Descartes had unsuccessfully tried to find Beeckman in Middleburg. Among the titles Matthias held were those of Holy Roman emperor and king of Bohemia. More than a year earlier, as his illness came on, Matthias had named as his successor his cousin, Archduke Ferdinand of Austria, and Ferdinand had begun acting aggressively, as if he already ruled. In Bohemia, he pressed his Catholic agenda on his new Protestant and Hussite subjects. They reacted in part by throwing some of his representatives out of a high window, the famous "defenestration of Prague" of May 1618. Conflict stirred but was constrained. Now, however, with the reigning emperor's death at last, the nobles of Europe expected that the Bohemian and Hungarian subjects of Ferdinand would rise up against him. All sides began to gather their forces. Within a month Descartes had heard of it: the active young cavalier was clearly in touch.

A few days later, Descartes took leave of his friend once again by letter, saying that he would be setting out for Denmark that day (April 29), expecting to spend some time in Copenhagen, the gateway to the Baltic. He included a quotation from Virgil's *Aeneid* about remaining uncertain "where fate may take me, where my foot may rest."[77] A Trojan setting out to see where destiny might take him, like Aeneas, the legendary founder of Tours. He hoped that Beeckman would send him a letter. Not long after, Beeckman would write to him at Copenhagen.[78]

And then the sieur du Perron slipped off into what would come to be called the Thirty Years' War. The next certain date we have is three years later, April 3, 1622, when he signed a document back in France, in Rennes, giving his brother Pierre power of attorney to sell some of the lands he had inherited (perhaps to pay debts incurred in the meantime). Three years is not, perhaps, a long time when looking back on a life, but it is everything for a young man who did not know what would happen next. Although his journey started well, René Descartes would be fortunate to make it out alive.

Baillet later explained that Descartes had not found a variety of occupations suitable to his talents, which he had promised himself on leaving France.[79] We might note that with the arrest and trial of Oldenbarnvelt and the upcoming conclusion of the Synod of Dort, and with the Spanish truce still holding, further action in the Dutch Republic would have seemed unlikely, whereas opportunity abounded elsewhere. According to this likelihood, the sieur du Perron remained a masterless man, looking for a field on which he could prove his worth in order to gain a patron. There would certainly be chances to become an Aeneas, or an Amadis.

Was he traveling alone? Knights-errant are rare. The letters to Beeckman say nothing about companions aside from a disparaging comment about the culture of the camp. But Descartes had learned to be discreet, telling Beeckman almost nothing about his personal life while making inquiries in the other direction, asking if Beeckman had yet found a wife. Descartes's chief biographer, Baillet, is silent on the subject. We do know from scattered references that at other times Descartes traveled accompanied by a personal servant at least and sometimes joined larger groups. He would certainly have had baggage to transport, and if he intended to fight, probably a warhorse or two as well as a riding horse. When he returned from the conflict in the east, he paid off most of the party then traveling in his employ and sold the horses. We might imagine him "on his own" if, despite a small entourage, he did not have companions of status similar to his own. But he may have had that sort of companion, too. In later years Descartes was known to have made an aborted journey to Copenhagen with Étienne de Villebressieu.[80] When Descartes was even younger, heading for strange places, it is easy to imagine that he traveled in company, perhaps with other young adventurers or perhaps among those who served a more senior figure.

Joining up with a group might also help explain his change of plans. For Baillet writes that in July 1619 Descartes left Breda and headed south. Baillet did not know of the letters to Beeckman, which are clear about how the noble volunteer set off for action via the Baltic. He was, however, insistent that Descartes ended up in Frankfurt on the Rhine, where he witnessed the formal coronation of Emperor Ferdinand on September 9; some years later Descartes, too, confirmed in a published comment that he had been there.[81] The likelihood, then, is that the sieur du

Perron had decided to make his fortune in war and begun his travels to the action via Denmark and the Baltic, as he informed Beeckman, but that he turned back to attend the coronation.

He would have had plenty of time to change course. Baillet's account gives us a few details of the latter parts of his travels. According to Baillet, on the way to Frankfurt Descartes stopped at Maastricht and then (not so far away) at Aachen, also known as Aix-la-Chapelle. In the latter place he learned of preparations being made for the coronation: since it was the old capital of Charlemagne, some of the treasures kept there were to be used in the ceremonial investiture at Frankfurt. From here on at least, the sieur du Perron was traveling with a group, perhaps even those who were escorting the imperial artifacts, for "they" next arrived (*etant arrivé*) in Mainz. There they confirmed that the local archbishop elector, Johann Schweikhard von Kronberg, had been summoned to cast his vote for the emperor.[82]

Mainz is near Frankfurt, easily connected by river. Baillet's account implies that on the last leg of the trip Descartes was traveling in the company of the party of the archbishop elector. The archbishop possessed ecclesiastical authority over a large number of bishoprics in the western parts of Germany and held temporal authority over territories on the Rhine and Main rivers, both above and below Frankfurt, where he was known as a Catholic moderate, supportive of Counter-Reformation institutions but not eager to attack Protestants. As an elector, he was one of the seven princes of the Holy Roman Empire who chose the emperor. We may imagine that the number of people who would have accompanied his delegation to Frankfurt would have been very large, so Descartes and any companions could readily have joined them for the remainder of the journey.

He therefore would have had more than enough time to begin his trip by going to Copenhagen and then turning around. What we have seen of Descartes so far suggests that as a young and energetic noble, he was making good use of his time. Even in winter camp, when sitting by the fire, and drinking, gambling, and whoring must have been common temptations for well-to-do young men, he was also improving himself by studying the local language and military engineering. If we look at his possible calendar, we know that Baillet says Descartes was at the coronation on September 9, and we can add that the elector of Mainz would have had to be in Frankfurt in time for the preceding vote, held on August 28. Let us suppose that the archbishop left Mainz no later than a week before the election. Working back, we can guess that the trip with a group from Aachen to Mainz could easily have been made by going overland to Cologne and then up the Rhine, probably easily done within a week. Before that, Descartes may have been traveling with only a small party and moving more quickly. Aachen could have been easily reached from cities on the Maas or Rhine; finding a return passage from Copenhagen to such ports as Amsterdam or Rotterdam, from which travel upriver would be easy to find, would not be difficult given the number and frequency of Dutch ships passing through the Danish Sound.

In other words, Baillet's account of Descartes setting off southward from The Netherlands in July and ending up in Frankfurt in plenty of time for the coronation is entirely plausible even if he undertook his first intention of going to Copenhagen and the Baltic before turning around. The timing allows him to have gone north at the end of April (as in his letter to Beeckman) and been back in the Dutch Republic again in July, heading toward Frankfurt. This scenario would have given him time to be present in Copenhagen and even regions beyond for a two-month visit or more, in a season when the days are long. Descartes had clearly intended to reach the scene of the impending conflict in Bohemia via the northern route, and he gives all appearance of having begun that journey, so we must ask what might have caused him to shift his attention to Frankfurt. Certainly, the importance of decisions to be made in Frankfurt would have drawn many eager young aristocrats. Perhaps Descartes simply decided that his personal interests lay in the possibilities of making contacts at the election and coronation. But he had been clear in his letter to Beeckman that in 1619 he considered his destiny to be bound up with reaching the fighting in Hungary. What might have intervened?

Perhaps Descartes simply changed his plans when he heard about the forthcoming coronation. At such a gathering there would be opportunities enough to seek a place among princely retainers. He had simply set out for Hungary on his own, ambitious for glory—many adventurers certainly did so—and then changed his mind after he saw new possibilities. But was he simply acting on his own? In light of Descartes's first comment to Beeckman about "not yet having been summoned" to the action, we need to consider the possibility that he set out in anticipation of such a summons. From whom?

Leading figures in France certainly had their own feelings of personal destiny, further stirred by recent events. It is well known that messianic and millenarian thinking pervaded the period; it is less noted in the English-language literature that the long struggle of France against the Habsburgs sometimes held out a vision of a Europe centered around the Most Christian King of France. In some of the circles Descartes had known in Paris, hopes abounded for a charismatic French king who would restore the world to an earlier age, an age before the worship of luxury or power, a pristine age of authentic rapport (*la sainte amitié*).[83] In other words, many longed for a Charlemagne-like figure to reign over an uncorrupted Christendom, returning the world to the old, sincere faith before the recent decades of doctrinal hairsplitting and hair-raising violence. Such views had underpinned the opinions of an earlier generation in France, including Guillaume Postel and Jean Bodin.[84] One of the prince de Condé's agents and secretaries, Isaac La Peyrère, would write about a universal French monarch who would reunite Europe and restore Palestine, building the new Jerusalem.[85] Cardinal Richelieu's adviser, the éminence gris Father Joseph, proposed a crusade led by the French king for the retaking of Jerusalem as a means to unify Christendom.[86] Duc Henri IV de Lorraine—closely related to the Guises from whom Descartes likely sought patronage—not only claimed the kingdom of Naples but also Sicily and Jerusalem by right of descent from the line of Anjou, and twice tried to seize Naples;[87] later, Descartes's patron Queen Christina of Sweden plotted to take Naples at the head of a French army to make it into a millenarian polity that would begin to reunite the world.[88] Christina's biographer, Susanna Åkerman, thinks that "her interest in Catholicism rested, paradoxically, on a political theory of universal concord and religious tolerance" that was "translated into a universal pan-Europeanism set out in the writings of a handful of Gallic prophets." She later wrote that Descartes had given her "the first lights" about Catholicism.[89] Descartes himself may have possessed grand ideas about the destiny of Europe, and shifts in the empire would have led to grand hopes among many, including figures in France.

Baillet also wrote that when he arrived in Frankfurt, Descartes was not able to attend the election of the emperor, only the coronation, because only those in the service of the electors were allowed into the city for the first occasion.[90] Baillet might simply be confirming that Descartes was present merely as a private person. But his precision might also suggest that Descartes had to wait on events because he was in service to someone who was not an elector. The possibility remains, then, that Descartes had originally aimed to get to Hungary not for personal glory so much as for aiding a great venture. If his father had introduced him to Philippe de Béthune, the comte de Selles—who had recently been working the conflict in the southern Alps between Savoy and Mantua[91]—such a link could have been active. He would later meet Béthune abroad, in Germany and Italy. Moreover, back in France a treaty had been negotiated between Louis XIII and his mother, Marie de Medici, for the moment reuniting the kingdom, so if Descartes had left Paris because of any scruples on that score, he might have felt newly enabled to serve Their Majesties. Or possibly the sieur du Perron continued to cultivate the patronage of the Guises of Lorraine. Lorraine's ruling duke was not an elector, but he was a prince of the empire, deeply concerned in events around the coronation. The opportunity to cultivate or renew ties with Lorraine might also have caused Descartes to change his plans and head for Frankfurt. All we know with reasonable certainty, however, is that an ambitious young Descartes, sensing his destiny and anticipating a call, intended to go to events in Hungary via the Baltic, got at least as far as Copenhagen, and then turned around to attend the coronation in Frankfurt.

The ceremony of the coronation took place on September 9 before a glittering crowd. Present were not only the seven imperial electors with their retinues, and the emperor-designate Ferdinand with his, but innumerable princes and princesses, dukes and duchesses, barons and baronesses, bishops and cardinals, the ambassadors of republics and kings, and many other great persons, along with their secretaries, clerks, and all kinds of members of their traveling courts. They in turn attracted countless other high-ranking people from all over Europe, together with those looking for favor, hosts of soldiers, churchmen and churchwomen, servants and retainers, and of course the countless numbers of charlatans, pickpockets, cooks, prostitutes, and vendors who flocked to any large gathering. In such a place, to pick up information and rumor about current events would have been as easy as simply listening in. Descartes knew some Dutch and was probably picking up other Germanic tongues; he had learned Latin, Greek, and Italian in school; many people spoke French as a common language. Descartes could have made his way through the babble. He later remarked that sometimes "we hear an utterance whose meaning we understand perfectly well, but afterwards we cannot say in what language it was spoken."[92] The only real question is whether he could have found lodging in the hubbub without losing his purse. But he undoubtedly enjoyed the pageantry of the coronation, which Baillet evokes in some detail, including the presentation of the sword, crown, and scepter of Charlemagne followed by the oaths and a *Te Deum*. Days of jousting and other public ceremonies followed. An attentive young nobleman would have had a lot of names to take in.[93]

It was also in Frankfurt that Descartes learned that Maximilian I, Duke of Bavaria, was seeking recruits. Maximilian would lead the campaign against rebel Bohemia. Maximilian also had connections with

Lorraine: in 1595, he had married his cousin Elizabeth Renata of Lorraine, daughter of Charles III, the ruling duke. Very possibly, then, Descartes's recent search for patronage from the duc de Guise had also established a bridge to Bavaria. Baillet insisted that Descartes was not in a position to worry about the rightness of any prince's cause, again hinting at the possible political implications that this move might have raised in the minds of his late seventeenth-century readers, when Brandenburg and other German princes were fighting Louis XIV.[94] He was also careful to avoid any implication that Descartes was acting against the interests of France by declaring that he acted as a volunteer.[95] But at the time, French policy was to support the Catholic princes of the empire, who were led by Bavaria, as a counter to the pro-Spanish faction at the Habsburg court in Vienna. As one notable historian remarked, "There was always a Bavarian party in France, and never was it more wide-awake than in the first half of the seventeenth century."[96] Dynastic ties between Bavaria and France also existed through Lorraine.[97] At the end of the year, Louis XIII himself would promise to aid his cousin, Emperor Ferdinand, at a moment when Bavaria fought for Ferdinand. So, at the time, in joining the party of the Bavarian duke, Descartes would certainly not be acting disloyally.

But time was pressing. The conflict in Bohemia and Hungary was already spreading into a more general conflagration. On August 26, almost contemporaneously with the election of Ferdinand as emperor, the Bohemians offered their crown to Frederick of the Palatinate. Frederick—only six months younger than Descartes—was not only one of the imperial electors and a powerful prince in his own right but also head of the empire's Protestant Union. He was related to most of the German princes, was a nephew of Prince Maurits, and had recently become the son-in-law of the king of England. Following the defenestration that began the revolt in Bohemia, Frederick had quietly supported an army under the command of Ernst von Mansfield that came to the aid of the rebels. (In Hungary, Gabriel Bethlen also launched a revolt against the Habsburgs at the same time, further weakening Ferdinand's position.) A united Bohemian Confederacy managed to depose Ferdinand and looked for an alternative, Protestant successor. The elector of Saxony declined the offer. The Bohemians turned to Frederick. He was reluctant to signal open conflict with Ferdinand, although in Frankfurt he was brave enough to lobby against him (the only one to do so). That isolated Frederick further, and most of his potential allies urged him not to seize the proffered crown. But on September 29 he decided his destiny otherwise and soon set out from Heidelberg for Prague, which he entered a month later. His coronation as king of Bohemia occurred at the beginning of November 1619. He opened negotiations with the Turks for assistance, and sought help from everyone else as well. It was going to be a difficult business.

For Emperor Ferdinand's part, while he had grave financial difficulties and other constraints, he also had an ally in Maximilian of Bavaria, who had been accumulating a treasury. On October 8, on his return to Vienna, Ferdinand stopped in Munich and signed a treaty that recognized Maximilian's leadership of a revived Catholic League and promised him both Bohemia and Frederick's electoral dignity as a reward for his assistance. The league itself met in Würzburg in December and agreed to raise an army with Maximilian at its head; he in turn delegated effective command to the Count of Tilly (an accomplished general who had learned his trade in the struggles of the low countries). Sending a shot across Frederick's bow, Ferdinand gave him until June 1, 1620, to surrender the crown of Bohemia. But the Protestant Union that Frederick headed had already agreed to raise men for its own defense, and in the spring of 1620 they, too, were on the move under the margrave of Ansbach. By June, the Catholic and Protestant armies faced each other at Ulm, on the Danube River. Descartes was there, having taken the duke's service but welcoming a delegation of French ambassadors who were trying to keep conflict from breaking out.

Anxious Dreams

Let us step back a moment, however, and note that during the winter of 1619–20 there was not yet an imperial army in the field for the duke to command. Baillet tells us almost nothing from that period about Descartes's relationship to events. It is as if Descartes went to ground after the coronation. As a kind of metaphor for this, instead of events, Baillet turns to a long description of three dreams Descartes experienced on the night of November 10, 1619. The account is based on entries in a notebook that has since disappeared but which Gottfried Wilhelm Leibniz saw and took notes on in turn, which confirm that Baillet reported it correctly.[98] In his later *Discours*, Descartes refers to the dreams briefly, framing them with a light touch in the following way: "I was returning to the army from the coronation of the Emperor," when "the onset of winter detained me in quarters where . . . I stayed all day shut up alone in a stove-heated room."[99] His dreams have since become famous and subject to much interpretative debate. But at the end of his report on the dreams, Baillet also drops a comment about how Descartes was afterward pulled back into events. There are some other clues about the winter of 1619–20, too. So what do we know?

The recorded content of the three dreams certainly suggests that Descartes was experiencing a state of high anxiety in which everything had to be seen afresh. Baillet's account begins by saying that the young man had decided to rid himself of all preconceived ideas, which was proving to be extremely difficult, even a torment.[100] He was in something like a state of shock: he had become exhausted, and he slipped into a kind of enthusiasm that left him open to dreams and passions. At the time, "enthusiasm" was often used as a term for people who reported having personal insights about divine intentions. Descartes was reading the signs, Baillet reports, seeking guidance. Baillet added a few pages later that Descartes had not been drinking even though it was Saint Martin's Eve, an occasion for revelry—even that he had not drunk a drop for a long time beforehand. Instead, Descartes thought that a spirit which was working in him had set his brain on fire and had anticipated these dreams. The first two dreams frightened him, while the third gave him a sense of promise.

The report of the dreams was clearly heavy with concern for the future, and Descartes explored their meaning. While later commentators usually treat it as a kind of vision on the road to Damascus induced by being closed up in an overheated room, the Frenchman might also have been in a state of considerable anxiety as a result of what would now be called culture shock, added to being exposed for the first time to the cold calculations of international power politics as viewed through the interests of a new but strange ally, Bavaria.

In his printed *Discours* Descartes says little about his dreams, instead discussing the conclusions he arrived at shortly thereafter, which sound much like an echo of Montaigne about the varying customs among people. Descartes had started to build his own views from scratch by first ridding himself of previous assumptions, he says. He comments, too, about being irritated by those who were always trying to reform the world, having himself learned that any opinion you might take had previously been advanced by one philosopher or another as definitive. On his travels, though, he had found that people who held views unlike his own were not "barbarians or savages." Fashions in ideas came and went just like fashion in clothing, which showed that "social custom and example" was the chief means of persuasion. He therefore developed four rules for coming to his own conclusions about the nature of the world, while also forming "a provisional moral code consisting of just three or four maxims." The first of these was "to obey the laws and customs of my country," in religion as well as in other fundamentals, and in all other things to act "according to the most moderate and least extreme opinions" of the place where he found himself. The second was to be as "firm and decisive in my actions" as possible; the third was "to try always to master myself rather than fortune"; and the fourth was to continue on his path of self-instruction. Adaptability, resolution, self-confidence, and keeping your eyes and ears open. After he had "established these maxims," he felt free to begin over again by speaking with other people in the real world, leaving his "stove-heated room" to set out on his travels once more.[101]

Descartes had after all been caught up in a whirlwind of events that was increasing in strength, blowing him who knows where, and the dreams spoke to that and resolved his fears. Indeed, the first dream began with whirlwinds, as well as a weakness on his right side (his sword hand). He was pushed by the winds to a school, tried to reach its chapel, and then was interrupted by an acquaintance who pointed him toward a certain gentleman in the courtyard who presented him with a melon from a foreign land. When he woke up he felt a physical pain and feared having been attacked by an evil spirit. He turned over onto his right side and prayed to God. After worrying about good and evil for a couple of hours, he fell asleep again, only to be awakened by his second dream, which took the form of a loud and sudden noise like thunder. When he opened his eyes, he saw sparks of fire floating in the room. He had had the experience before, but this time he opened and closed his eyes several times and his terrors floated away.

In the third dream, two books appeared to him and then a figure became manifest, who handed him a verse later explained to be from Pythagoras; as they entered into discussion, the content of the books started to shift so that Descartes could not find the verses for which he was looking; the books and the figure then disappeared. Not being frightened awake this time, he slept on and managed to find an interpretation for the dream while still asleep, deciding that it indicated the union of philosophy and wisdom. The poets, he decided, were full of maxims more succinct and powerful than anything offered by the philosophers, speaking from enthusiasm and imagination, like flints producing sparks. So he awoke and decided that the "Spirit of Truth" had come to open to him "the treasures of all the sciences" (meaning all knowledge). He had seen in the last dream some copperplate engravings, but after receiving a visit from an Italian painter the next day, he sought no further explanation.[102] He was now in a world of strange events, but it all augured well. He had received a message about trusting his creative intuitions that also warned against trying to force himself into sacred spaces.

On the face of it, it is not surprising to find the young nobleman feeling powerful winds pushing him about, nor perhaps feeling that they were keeping him from entering consecrated places. Baillet says that Descartes interpreted the melon as "the charms of solitude" that humans ask for and the wind as an evil spirit trying to push him into a place from which God (Baillet's word) held him back. The explosive dream was "the Spirit of Truth descending to take possession of him." His third dream confirmed that truth emerged from the aphoristic spirit of poets, from their intuition about reality rather than from reasoning, which gave him self-confidence. He prayed in thanks and promised to make a pilgrimage before the end of November to the shrine of the Virgin Mary in Loreto, walking the last part all the way from Venice. But, Baillet writes, he had to postpone the trip "for reasons that remain wholly unknown." He then spent the winter composing a treatise and by February he was beginning to seek a publisher.

It may be, then, that the "dreams" were in fact a sketch of what we might now call a literary work similar to what his friends had been reading and discussing during his Parisian period but reflecting his new appreciation for strangeness. His words include phrases such as *enthousiasmus*, meaning a state of divine possession that the ancients associated with dreams; *force de l'imagination*, which is the power to strike an image that links the inferior and superior worlds; and *génie* for the Latin *spiritus*, a word associated with the presence of djinns and the spirit of truth.[103] The report of the dreams also includes a verse from Pythagoras (who had himself reported on how dreams lead to wisdom), while other verses that appeared and disappeared were thought to be indications of the union of philosophy and wisdom in the work of poets. At the time, well-known literary exemplars of powerful dream sequences ranged from Cicero's "Dream of Scipio" to famous reported revelations of Saint Paul and Saint Augustine, while in Descartes's own day, dream interpretation remained an important phenomenon, as in the visions experienced by Ignatius Loyola, founder of the Society of Jesus that had educated him. In 1608, the Lutheran Johannes Kepler, too, wrote about his new astronomical science in the form of a dream, although his *Somnium*, sometimes considered an early work of science fiction, would not be published until 1634. Another of Descartes's contemporaries, the chemical physician Jan Baptista Van Helmont, a Catholic accused of dangerous heterodoxy by the Inquisition, also recorded dreams as a mark of receiving divine enlightenment, dreams that are sometimes compared to those of Descartes.[104]

There is likely to be another source for much of the content of the recorded dreams, if not their form, one of the most important works of skepticism in the period: Pierre Charron's work on moral philosophy, *De la sagesse* (*On Wisdom*, first published in 1601, two years before his death). In the 1990s, Frédéric de Buzon was looking through a book in the city of Neuburg and noticed Descartes's name in an inscription on the flyleaf. It read: "To my very learned friend and little brother" (an expression of familiarity) "René Descartes, given in friendship, father Johannes B. Molitor, Society of Jesus, done in the year 1619, [signed] JBM."[105] Buzon confirmed that Molitor was a Jesuit but could not confirm where he was assigned at the time, although he might well have been at Neuburg, where a Jesuit house was then under construction. The book itself lacked the title page but can be identified as Charron's, probably the third edition, of Paris 1607.

Charron was a priest who served Marguerite de Valois, a daughter and sister of kings and first wife of the future Henri IV, a powerful woman with many lovers and associates. Charron also became a friend of Montaigne and was given his coat of arms to wear after his friend's death. In presenting his views on Christian skepticism, Charron drew heavily on his more famous friend's views, although Charron's work was later attacked by the Jesuit François Garasse as atheistical, even one of the main touchstones of the party of atheists. But that was yet to come. Perhaps Molitor left his young friend the book in order to get Descartes's opinion of it. Or perhaps he was eager to get the learned nobleman's help in dissecting it for criticism. But we can be certain that he presented it as a gift, something that he expected to be appreciated by its recipient. If Descartes did not know it already, he probably saw the book's arguments as entirely in keeping with what he had encountered a few years earlier in Paris.

Recent historians of philosophy have noticed many close parallels between Charron and Descartes. Charron aimed at finding wisdom through judgment, but to get there he began by showing that as mere mortals we humans do not have access to certainty, only to likelihood (*vraisemblable*). He attacked all kinds of superstitions and dogmatisms based on hypotheses or pretended access to truth, dividing the path toward knowledge, which is rooted in the world, from theology. Ethics could not be subordinated to religion, then, nor did it address such questions as sin and grace, nor personal salvation. Institutional religion was a political embodiment of local tradition. A wise man would therefore act according to his conscience but remain publicly silent and masked on the fundamental questions of universal truths.[106] Even without certainty, however, the wise—a small group among humankind—could nevertheless achieve a good life, although on the way there one needed to be guided by some provisional rules of morality. These turn out to be almost identical to Descartes's own rules. Descartes's dream episode with the book of poetry invokes, in part, Charron's "yes and no" (*oui et non*), which appears on the missing frontispiece of the copy in question.[107] (Did Descartes tear it out to keep when he left the book behind?) In Descartes's own later account of the episode, the dreams are followed by his moral rules for navigating the world. Descartes had wanted to go into the chapel in the school, and had even first been pushed in that direction by the whirlwind but was then kept from entering, finally being diverted by an acquaintance to speak with a man from a foreign country. He offered him a melon. But Descartes left the book behind.

Curious Meetings

According to Baillet, within a few days, "he got quit of his Enthusiasm,"[108] for his further thoughts were interrupted.[109] Of course he was interrupted: he had joined the duke, who was mobilizing his forces, and the sieur du Perron would have had duties to perform.[110]

Recent biographers have debated whether Descartes spent the winter in Ulm or Neuburg. Both are on the Danube, and strategic crossroads; both were on the borders of Bavaria. But Ulm had joined in the Reformation in 1530 and became well known for possessing the largest Protestant cathedral in Europe, while Neuburg was the capital of the duchy of Palatinate-Neuburg. Many biographers prefer Ulm as Descartes's winter quarters because of its association with the mathematician Johannes Faulhaber, whom Baillet says Descartes met. On top of its city walls small dwellings still remain that once were built as soldiers' residences: it is a pleasant thought to imagine that in one of them Descartes had his dreams. But at Neuburg, Count Wolfgang Wilhelm had converted to Catholicism a few years earlier, had recently dedicated a church to the Virgin Mary—whom Descartes had promised a pilgrimage—and invited the Jesuits to take up residence. For these reasons, Geneviève Rodis-Lewis prefers to think that Neuburg is where Descartes spent the next few months.[111] Neuburg is certainly where Descartes was given a copy of Charron's book.[112]

Moreover, Count Wolfgang Wilhelm of Neuburg descended from the Wittelsbach dynasty and married Magdalene of Bavaria, sister of (Wittelsbach) Maximilian of Bavaria and daughter of Renata of Lorraine. As the head of the cadet branch of the Palatinate line, the count was also a relative of Frederick, the Elector Palatine, whose daughter, Princess Elizabeth, Descartes would later serve. He also controlled Jülich, on the Rhine, where he had been supported by the French king Henri IV on the eve of his assassination and where he would have the support of the Spanish army in the low countries. Two years later, on his return to France, Descartes attended the siege of Jülich until Ambrogio Spinola took it. Many years later, Descartes famously attended the court of Queen Christina of Sweden, whose female ancestors were also Wittelsbachs. It is therefore likely that through a connection to Lorraine, Descartes had entered into service to Maximilian of Bavaria and might even have established a relationship with the court of Neuburg.

But we do not have to imagine that Descartes simply went to ground for the winter in one place or another. Armies often had to do so, since moving and provisioning large numbers of men and their followers, and transporting their cannon and baggage, became difficult at certain times of the year. But individuals and small groups could get about year-round, only occasionally being blocked by snow in the high passes or a blizzard. As we have seen, Descartes was often on the move. What he recounted in the *Discours* supports the impression of activity, for although he reported that for a short time it was cold enough to stay enclosed in a room with a stove, several pages later he picked up the story by writing that "I set out on my travels again before the end of winter. Throughout the following nine years I did nothing but roam about the world, trying to be a spectator rather than an actor in all the comedies that are played out there."[113] Note that his travels began "before the end of winter": he started on real journeys even before the worst of the cold was past. And while his word *travels* implies voyaging from one place to another, he might earlier have been making short trips that brought him back to a home base. Why should we imagine that he was cooped up in a room for more than half a year, until the French ambassadors arrived at Ulm in June?

Baillet says that Descartes had to postpone his pilgrimage to the shrine of the Virgin in Loreto "for reasons that remain wholly unknown." Perhaps Descartes first assumed that he would be going to Venice on business for the Duke of Bavaria. The Venetians had waged a war against Archduke Ferdinand with Dutch and English help, which had ended in 1618; their next move would be important. Or Descartes may have had other reasons to think that he would be off there, unconnected with the duke. But the leaders of the Catholic League decided to gather in December in Würzburg to discuss raising an army. Würzburg is not far north from either Ulm or Neuburg. Other delegates got there and back from all over the German lands despite the season. We might guess that the reasons unknown to Baillet for not being able to get to Loreto included attending on one of the dukes in Würzburg.

A hint from the archives suggests that Descartes was on the road. In this case, he was carrying dispatches from Ulm about the time that Baillet says he had to abandon his writing project, in February 1620. According to Édouard Mehl, in 1972, L. Gäbe noted that the rector of the lyceum in Ulm, Jean-Baptiste Hebenstreit, wrote a letter to Kepler dated February 1, 1620, in which he asked if he had received a packet of previous letters carried by a certain "Cartelius." Kepler resided in Linz—like Ulm, also on the Danube, although east of Munich. (Kepler himself would later move to Ulm.) If this is our man, then Descartes must have set out for Linz in late January.[114]

We can be reasonably certain that Descartes was already keen on optics, having met Claude Mydorge in Paris, and he later based much of his published work on Kepler's principles.[115] One recent analyst thinks that Descartes's unfinished writing project may have been an early manuscript on corpuscularian optics prompted by reading Kepler.[116] In the winter of 1619–20, however, Kepler's interests had moved on: not only was he mounting a defense of his mother against charges of witchcraft, but he also had just published his work on the "Harmony of the Worlds" (*Harmonices Mundi*, 1619), which illustrates the relationships between geometrical and physical knowledge; and he continued to work on his *Epitome astronomiae Copernicanae*, the most up-to-date explanation of the motions of the heavens based on the assumption that the earth revolved around the sun while also spinning on its own axis.[117] (Volume 1 had appeared in 1618, and volume 2 was about to be published, in 1620; the third and final volume would be completed in 1622).[118] We cannot be certain whether Descartes took an interest in mathematical astronomy at the time, but it certainly was a subject being widely debated. The appearance of three comets during 1618–19 had led to public controversy not only about their astrological significance but also about their nature, especially whether their physical appearance in the heavens beyond the sphere of the moon was evidence for heliocentrism. Galileo was among those who took the opportunity to press the new astronomy forward on the basis of the comets.[119] Descartes was later clearly committed to heliocentrism. He could well have been encouraged in such views by Beeckman—also a Copernican—but his probable meeting with Kepler might have solidified his opinion.[120]

According to Baillet, Descartes also sought out the mathematicians Faulhaber and Peter Rothen. Faulhaber was in Ulm, and Baillet places the meeting there some months later, about the time of the arrival of the French ambassadors. He reports that Descartes was trying to find competent mathematicians, and so he paid a call on Faulhaber.

Faulhaber received him civilly but mistook him for a simple soldier and insulted him about his mathematical abilities, at which Descartes bridled. He responded with a challenge, answering every question that Faulhaber could put to him, after which Faulhaber invited him into his rooms and showed him a recent book he had written. Descartes looked through it and not only solved all the problems it set but also explained the basis for their solution, of which Faulhaber was ignorant (according to Baillet). A similar set of circumstances occurred when in the same period Descartes in Nuremberg visited Rothen, who had proposed solutions to several of Faulhaber's questions and had Faulhaber solve several of his in turn.[121]

Was Descartes seeking out the Rosicrucian brotherhood, as is sometimes thought?[122] Faulhaber had dedicated his 1615 *Mysterium arithmeticum* to the group, and it contains a mathematical table able to produce polygonal and pyramidal numbers that would yield the special numbers in the Bible, allowing further progression in decoding apocalyptic chronology.[123] According to Susanne Åkerman, the work was dedicated to "Polybius Cosmopolita." Also known from notes taken by Leibniz on Descartes's notebook (the "Cogitationes Privatae"), is that in 1619 Descartes had developed a plan to write a book under the pseudonym Polybios de Cosmopoliet, titled *Thesaurus Mathematicus*. It would be dedicated to all the erudites, but especially to the Rosicrucians of Germany, and it would solve any mathematical problem according to the various methods set out in it. Some scholars therefore consider the title to be not so much a plan for a book as Descartes's own little joke aimed at Faulhaber and his peers.[124] Perhaps the proposal for the *Thesaurus* (like the dream sequence near it) was a literary exercise, a fragment "de l'Histoire de vostre Esprit" that Balzac and his friends later anticipated from

Descartes, which would contain "your various adventures in the world and in the higher regions of the atmosphere" (*vos diverses aventures dans la moyenne et dans la plus haute region de l'air*), suggesting a microcosmic/macrocosmic view not unlike that in many of the tracts associated with the Rosicrucians.[125] Whether meant as a plan for a serious publication or a private joke, it is plain that Descartes's note about the *Thesaurus* was a response to his meeting with Faulhaber.

Books attributed to the brotherhood had been appearing recently, with the first two printed in Kassel in 1614.[126] They promised the revelation of a more perfect knowledge of Christ and of Nature, as discovered by the illuminated founder of the brotherhood, Christian Rosencreutz. The *Fama* (1614) set out the history of Rosencreutz, who gained a knowledge of universal truths after voyaging to Damascus, Damcar, and Fez, and finding the book *M*, which he translated into Latin. He went on to found a society of eight chaste knights who possessed the occult knowledge contained in it and who would have their "third succession" 120 years after Rosencreutz's death, when his secret tomb would be discovered in the building he had constructed, the House of the Holy Spirit. The *Confessio* (1615) elaborated on Rosencreutz's promise and gave his life dates as 1378 to 1484, placing the kairos year of the discovery of the tomb and the general reformation at 1604.[127] That was about the time that the manuscripts were first written, and they circulated for a decade before appearing in print, claiming to contain the possibility of a universal reformation in religion, politics, knowledge, and social relations stemming from the original revelation of God to Adam and his (authentic) successors.[128]

In the moment of political anxiety associated with the outbreak of war, such works were also read as prefiguring current events.[129] They were among several documents apparently first composed in a Lutheran circle around Tobias Hess and put on paper by Johann Valentin Andreae, perhaps lightheartedly conveying some serious views about nature that were widely shared among alchemical and Paracelsian adepts by linking the microcosm and macrocosm with the centrality of Christ.[130] (There were many important practitioners of Paracelsian alchemy—or chymistry—in the German lands, and they often proposed a kind of immaterial corpuscularism based upon the three principles of mercury, sulfur, and salt.[131]) Having then been condemned as a student for immoral (libertine) behavior—there were certainly many associations between libertinism and other cosmologies that fed Rosicrucianism[132]—Andrea spent the years between 1607 and 1614 traveling in German, French, and Italian cities before accepting a position as deacon in a church near Stuttgart, and in 1620 in the Black Forest town of Calw.[133] Some decades later, two English virtuosi who took an interest in these matters, Robert Child and Elias Ashmole, thought that the Rosicrucian brethren lived in a monastery seven miles from Strasbourg, which is just on the edge of the Black Forest, in turn associated with many of the early figures of the story.[134] Ulm is a bit further east, in Swabia, a land associated with the legend of Faust.

Faulhaber was not alone in publishing books dedicated to the brethren in an attempt to attract their interest. Other mathematicians with whom Descartes was in touch also expressed their interest in decoding the universe: Kepler, for instance, was keen to explore the musical/mathematical harmonies of the spheres.[135] Descartes himself had expressed an interest: when he first set out from the Dutch Republic, Descartes stopped in Dordrecht, an important site on the great inland rivers of the low countries from where one could easily find shipping onward to almost anywhere. At an inn there, on April 26, he had a long conversation with “a learned man” about his method of using Raymond Lull’s *Ars parva* (*Smaller Art*)—a famous millenarian work[136]—to invent a speech of more than an hour on any topic, even to speak for twenty-four hours if pressed. The man further elaborated on Lull and another mystic, Cornelius Agrippa, explaining the “keys” they had left out of their books, which the man claimed to have discovered. In cryptography of the time, keys were similar to passwords, allowing decryption of encoded messages: keys unlocked deep secrets. Descartes knew that Beeckman had a copy of Lull’s book and invited him to look into it.

Given the report of Descartes's earlier interests, his recorded dreams, and his own sense of having discovered a kind of universal science based on mathematics, it seems entirely possible that he was seriously interested in finding out about the Rosicrucians. Both Leibniz and Huygens later criticized Descartes for his mysticism.[137] In another set of Descartes's early notes, the "Studium bonae mentis," he referred to the existence of the brethren as uncertain—*necdum . . . quidquam certi*—but that is not a simple dismissal.[138] Faulhaber may have included Descartes in his own plans.[139] Not surprisingly, then, some historians have been tempted to take seriously the possibility that Descartes shared with the brethren (and the poets) a sense that true knowledge can transform the world for the better, and that the light of inspiration is a source of such knowledge. Descartes may well have been moving among mathematical adepts who shared their secrets with one another, and the hermetic view that "instantaneous activity signifies creation" may have been very important to the development of Descartes's theories of light and corpuscular matter.[140]

But there was also a more immediate reason for Descartes to seek out Faulhaber and Rothen, and possibly Kepler as well: practical mathematics. If Descartes had joined the duke's "army" when Baillet says he did, he would have been something like a staff officer at a moment when help was needed for gathering the forces. While Maximilian might have welcomed people like Descartes as soldiers, he may have been looking for other kinds of assistance, too. If we think again of the example of the story of Viète, for example, he simply turned up when Henri IV asked for him because he was present with the king at Fontainebleau. Princes required mathematicians to be near, for cryptography as well as for engineering. Faulhaber was not simply a mathematician and a Rosicrucian sympathizer, he was also the city surveyor of Ulm, well known for his ability to design fortifications. He also designed fine mathematical instruments for the use of military engineers. In 1628 Kepler himself took up service with the imperial general, Wallenstein. The account of Descartes meeting Faulhaber and Rothen was first reported by Lipstorp,

Descartes's first biographer and the most concerned to underscore Descartes's military service.[141] Descartes may well simply have been charged with recruiting other people like himself who could serve the duke as mathematical experts. Faulhaber at first took him to be a plain soldier but soon became interested. Plans for battle were coming together.

*

By the end of the winter of 1619–20, then, Descartes had become a well-educated military engineer who understood the principles of the art of war. He had gone looking for his destiny, in all likelihood in association with Wittelsbach-Lorraine interests in the Holy Roman Empire, but perhaps also feeding information back to French diplomats. He had certainly gone through a rite of passage, experiencing his whole world coming unstuck but also taking heart from his dreams or imagined dreams. The process had led him to discover a few basic rules to live by in foreign places. He was also attending to the business of the mobilization for war. It would be a big one. The whirlpool tugged at him. He would be lucky to come out into the air again.

PART 4

War and Diplomacy in Europe

For the next five years, from the spring of 1620 to the spring of 1625, Descartes's itinerary shows him to be in some of the most critical places for French interests in central Europe. For the first two years, he was bound up with the excitement and horror of war in Bohemia and Hungary, where he almost lost his life in a major military disaster. Then he was in Poland and the German-speaking Baltic, then after a brief return to France in the region of the southern Alps known as the Valtellina, and in Rome, just when France was beginning its Italian campaign. He made many further acquaintances, including members of the Francophilic Barberini family, one of whom sat on the papal throne. His visit to Italy also happened to coincide with the years when Galileo was making a further stir with his book called *The Assayer*, and there Descartes's interest in natural philosophy would revive.

But mainly his late twenties were spent moving between battlefield and negotiating table, an active participant in major events even if not one of the notable names. He might be thought to be a volunteer member of the army and diplomatic corps, although fortune would take him in yet other directions. He later underlined those activities in his life at the time in his *Discours*, although in a veiled way, writing that during this period of his life he was "travelling, visiting courts and armies, mixing with people of diverse temperaments and ranks, gathering various experiences, testing myself in the situation which fortune offered me, and at all times reflecting upon whatever came my way so as derive profit from it."[1] Descartes was certainly becoming wise in the ways of the world.

Into Bohemia

No later than the spring of 1620, Descartes became aware of a delegation of ambassadors from France headed for Ulm. For a generation, France had seen itself as attempting to keep the peace in Europe, well knowing that interlinked personal and political interests meant that a single major conflict could well lead to a general conflagration. But the new king Louis XIII allowed the promptings of his religious conscience to guide his actions, promising to send financial support to the new emperor Ferdinand to put down the rebellion in Bohemia. When that proved to be fiscally impossible for him, he sent ambassadors instead.[2]

The French delegation crossed into Germany in the spring and reached Ulm on June 6, 1620. According to the seventeenth-century biographer Adrien Baillet, Descartes rushed to greet them, "several of whom were of his acquaintance."[3] Indeed, the leaders of the party included the duc d'Angoulême, one of Descartes's possible patrons from a couple of years earlier. Another was Charles de L'Aubespine, marquis de Châteauneuf, who had previously served as French ambassador to the United Provinces and the Spanish Netherlands—in later years he would conspire with the duchesse de Chevreuse. The third senior diplomat may also have known Descartes: Philippe de Béthune, comte de Selles, who could have been introduced to Descartes by his father in the weeks between René's exit from Paris and his departure for The Netherlands. Descartes might also have made a connection to Béthune through Balzac, for Béthune had been among those negotiating with Marie de Medici on behalf of her son Louis XIII, at the Balzac house in Angoulême, in 1619.[4] If they had not met previously, Béthune may well have

Map 1. Places associated with Descartes in Europe. Drawn by Lyse Messmer.

been interested in Descartes now that he had experience abroad; they would later meet again in Rome.

Within a few weeks, Baillet notes, the ambassadors had attracted so many young seigneurs and other persons like Descartes that the French counted more than four hundred horses in their suite.[5] They were aiming to arrive at a potential flashpoint before the outbreak of hostilities. At Ulm the army of the Protestant Union, about thirteen thousand strong, faced about thirty thousand troops of the Catholic League massing downriver.[6] Angoulême's party initiated talks between the two sides and on July 3 had obtained their signatures on a nonaggression treaty. Peace at Ulm gave the advantage to Maximilian I, the Duke of Bavaria, however. The army of the Protestant Union under the margrave of Ansbach was now free to move against the Spanish forces that had invaded the Rhenish Palatine under Ambrogio Spinola, but did so ineffectively, while the duke's own and larger army was now free to restore the emperor's power in Upper Austria (they would enter Johannes Kepler's Linz at the beginning of August). Maximilian's general, Tilly, then headed further east to threaten rebellious Bohemia. At the same time, an imperial army under Charles Bonaventure de Longueval, Count of Bucquoy—another veteran of the wars in the low countries—headed from Vienna into lower Austria to restore Ferdinand's authority there. First jobs done, the imperial armies of Tilly and Bucquoy joined up and crossed into Bohemian lands. Events would roll toward the Battle of White Mountain, usually taken as the opening act in what became the Thirty Years' War.

For this part of his biography, Baillet was presumably working from the now lost account of Descartes's memoirs of his time serving in the wars. He seems not to have known for certain whether Descartes left Ulm with the Catholic army under Tilly or whether he stayed with the French ambassadors and their train, but he thinks it was more likely to have been the latter case.[7] Unlike more heavily urbanized lands, in the large spaces of central Europe armed forces could seize ground, cut off approaches, go around, and otherwise maneuver, making prolonged sieges less necessary. A minor French noble attached to the Duke of Bavaria's party might provide better service by remaining with the diplomats and staying alert to events, or even serving as an intermediary, than taking up space among the horses, kitchens, and baggage of a commander's entourage. The duc d'Angoulême's party made their way down the Danube in the hope of negotiating a treaty between Emperor Ferdinand and King Frederick of Bohemia, arriving in Vienna less than three weeks after the treaty of Ulm, on July 20. Having just freed Maximilian's army for fighting the emperor's enemies in the east, however, their offers of further diplomatic assistance yielded little response from Ferdinand.

According to Baillet, Descartes then joined the ranks of the young Frenchmen surrounding the comte du Bucquoy. Bucquoy had earned an impressive military reputation in the army of Flanders, commanding the Spanish artillery, and had been ambassador extraordinary to France in 1610; he was much liked by many of that nation.[8] Baillet insists, however—perhaps thinking of how Descartes later served the family of

Frederick—that the young man did not fight so much as he observed affairs. Perhaps he continued to serve as an intermediary on behalf of the Duke of Bavaria, to whom he would soon return. In any case, there was certainly much to notice. While the Bohemian forces were mutinous because of a lack of pay, Frederick found ways of resisting nevertheless; on the other side, the imperial army started to die in large numbers from the "Hungarian fever"; and regulars and irregulars everywhere resorted to the most horrible acts of violence against the locals.[9] As autumn took hold and maneuvering in the field continued, cold and hunger added to the misery.

The final confrontation occurred on November 8, 1620, at Bílé hoře, White Mountain, on the outskirts of Prague. Frederick seems to have been unprepared and his troops were forced into a retreat, which turned into a rout. On the next day, Frederick and his queen, Elizabeth Stuart, fled Prague with their family and retainers. The city was put to the sack, temporarily satiating the imperial soldiers. When news of the event first arrived in Vienna, the French negotiators were so surprised that they awaited confirmation to come from other couriers. They accepted thanks for their unsuccessful efforts in negotiating a peace but stayed on, now focusing on the emperor's relations with the Prince of Transylvania, Gabriel Bethlen, a vassal of the Ottoman sultan who was trying to revive the Kingdom of Hungary.[10] (The French ambassador in Constantinople had made contact with Bethlen in 1619, but the ambassadors in Vienna had little leverage with him, so in 1621 the embassy was recalled.[11]) According to Baillet, Descartes had been in the field during all these events and contributed to the victory at Prague, still retaining his rank as a volunteer with the Duke of Bavaria. He remained in the despoiled city with the occupying forces for six weeks before campaigning with some of the Bavarian troops into the far south of Bohemia; then he returned to the Duke of Bavaria's Munich.[12]

It had been November, with winter coming on. If the imperial army had not headed straight for Prague but more cautiously probed Frederick's forces, or if Frederick and Prague had been ready to sustain a siege, or if any other of a dozen possibilities had transpired, the imperial army, shrinking already from disease, would have had to go into winter quarters demoralized. Perhaps that would have given Frederick's potential allies a chance to come to his aid. Perhaps the mooted alliance of Bohemia, France, Venice, Savoy, England, and The Netherlands might actually have emerged as a third force in Europe to calm the divisions between Tridentine Catholicism and militant Protestantism. Only a year earlier, Paolo Sarpi, a friend of Galileo, had allowed his *Istoria del Concilio Tridentino* (*History of the Council of Trent*) to be published in London, and editions in Latin, English, French, and German had followed shortly, showing the doctrines of the church not to be infallible pronouncements of truth but negotiated deals struck by international politicking by a princely pope and a centralizing curia. There was talk of unifying Christendom in other ways, without the sword. But also in 1619 the Prince of Orange had the moderate Johan van Oldenbarnevelt's head removed from his neck, while elsewhere imperial forces took the field to put down any signs of disloyalty to a Catholic emperor. For a moment the world had paused, unsure of which way events would move, with peace a real possibility. But the surprisingly complete victory for the emperor's soldiers at White Mountain confirmed that other avenues were now closed. Ideology and armies would continue to take precedence over *politique* negotiation.

Descartes's early biographers had no doubt that he took a personal part in this change in the fortunes of Europe.[13] That means he had served the causes of two bitterly opposed princes, one Calvinist and one Catholic, each using the rationales of dogma to support their power. Yet he seems to have been relatively indifferent to such matters, placing his faith in premonitory dreams and the wisdom of poets. The political morality of the personal and ideological clashes behind the firestorms of Europe were for the greats to decide; he seems simply to have served as intuition directed.

To Hungary and Disaster

Descartes remained with the Duke of Bavaria's forces until the end of March 1621, when he again left the duke's army for the Count of Bucquoy's, then in Hungary. From this point forward, Baillet's account becomes more detailed, either because he was not required to be the soul of discretion or because the materials he was working from contained more information. Descartes was now definitely involved in the action. Bethlen had refused the terms he was offered by the French negotiators and fought on against the emperor. Bucquoy moved against him first by laying siege to Pressburg/Bratislava in early May, where Descartes saw action. Successful there, the count attacked Tyrnau/Nagyzombat, which did not resist long, and continued east toward Budapest, seizing cities and towns along the way. Descartes acquitted himself well in all these actions, Baillet says; with seizing cities now the main method of attack it was the kind of campaign where officers who knew something about military engineering would be useful.[14]

But a problem arose at Neuhäsel/Nové Zámky, which brought Descartes face-to-face with disaster. The determined city managed to hold out against Bucquoy's troops for six weeks. Early in July, in support of those resisting in the city, Bethlen's forces began to arrive in the field in number, besieging the besiegers, who became trapped between the town's bastions and the Hungarian cavalry ranging the countryside. With his own troops and horses experiencing hunger, the Count of Bucquoy led a foraging expedition in force on July 10, encountered strong opposition, and in the ensuing fight was killed. Demoralized, and with the rest of Bethlen's army coming on the scene, on the next night the imperial army lifted the siege and tried to retreat across the Neutra River. The marshy ground on either side made it almost impossible to cross. It must have been a special nightmare for the engineers, trying to save what they could of their artillery and equipment. Less than half the number of men with which Bucquoy had begun the campaign—about eight thousand—made it alive onto Schütt Island in the midst of the river, where they held out for some time. Descartes himself may have thought himself wounded, only to find that after the heat of battle he had felt "a buckle or strap caught under his armor, which was pressing on him and causing his discomfort."[15] Having turned events around, Bethlen went on the offensive. He recaptured Tyrnau on July 30 and from August 18 onward besieged Pressburg, which had been Bucquoy's first success of the year.[16] A truce would be arranged in October. Descartes was now limping through a military collapse, his commander dead, facing defeat and in danger of his life.

Descartes later referred to military action as an example of how he learned to respond to the passions that sometimes overwhelm one in the moment. "For example, when we are unexpectedly attacked by an enemy, the situation allows no time for deliberation; and yet, I think, those who are accustomed to reflecting upon their actions can always do something in this situation. That is, when they feel themselves in the grip of fear they will try to turn their mind from consideration of the danger by thinking about the reasons why there is much more security and honour in resistance than in flight. On the other hand, when they feel that the desire for vengeance and anger is impelling them to run thoughtlessly toward their assailants, they will remember to think that it is unwise to lose one's own life when it can be saved without dishonour, and that if a match is very unequal it is better to beat an honourable retreat or ask quarter than stupidly to expose oneself to a certain death."[17]

Baillet put a good face on things, but Descartes had clearly seen enough. Baillet reports the imperial troops retiring in good order after the death of Bucquoy—meaning only, one supposes, that the defeat was not a rout—with Descartes reaching Pressburg on the night of July 27 in the company of French and Walloon troops (the latter being French-speaking soldiers from the low countries). One can infer, however, that Descartes must have made it to Pressburg with a group of soldiers who had either been fighting rearguard actions in an organized retreat—never a pleasant business—or who had spent some days on the move by night, perhaps from Schütt Island, slipping through territory controlled by enemy cavalry, attempting to make their way to the main imperial stronghold where a last stand was being organized (and which would hold). He would later knowingly write about how it takes greater skill for military leaders "to maintain their position after losing a battle" than "to take towns and provinces after winning one."[18] Baillet would comment that Descartes had become disgusted with the profession of arms (*acheva de la dégoûter de la profession des armes*). The volunteer had lost the commander he admired and probably tried to impress, leaving him stranded far from France without chance of advancement and even in mortal danger. Baillet probably meant not that Descartes had simply lost his stomach for war, since for some years yet he would stay close to military events, but that he had given up on making his personal mark on the battlefield. It was no romance of knights-errant, or even the straightforward application of practical mathematics, but a dangerous and unpredictable business.[19]

The Baltic and Return to The Netherlands

Descartes seems to have decided to go far and fast. Whether he did so in the company of others is again unknown but likely. Baillet explained that he wished to travel because he hoped to see a variety of people and customs and to study the grand book of the world (*le grand livre du monde*, a phrase Baillet lifted from the *Discours*, although it might originally stem from Montaigne). Baillet says that Descartes did not wish to return to Paris because of internal conflicts there and an epidemic of plague, deciding instead to see parts north. So via either Poland or the Ottoman breadbasket of Moldavia he headed into Silesia, stopping at Breslau/Wrocław (which had just capitulated to the emperor). Baillet denies the rumor that Descartes fought for the Turks because (he thought) the Polish–Turkish war ended at about this time, although in fact Polish resistance proved stronger than expected and the conflict continued into October.[20] Military activity in the region certainly continued to make travel difficult—another reason to think that Descartes traveled in company. No one could accuse him of desertion, Baillet continues, since he had joined as a volunteer and, one might add, the commander to whom he had pledged loyalty was dead. But he had now given up serving the emperor in arms.[21] Baillet then has the sieur du Perron continuing north to Pomerania on the Baltic coast, which he reached by early autumn.

Could Descartes have been acting on behalf of someone? He could already have been in the region in the spring of 1619. Moreover, on the southern shores of the Baltic, Poland, Sweden, Denmark, and several German princes were contending for dominion, with occasional intervention by the Dutch, whose economy depended heavily on their control of the Baltic trade. The Danes had recently imposed a peace on Sweden that forced the latter to pay war reparations; Christian IV of Denmark was investing the income aggressively, as in building the city and fortress at Glückstadt downriver from Hamburg, which resulted in forcing that great city to recognize his suzerainty.[22] Sweden and Poland had entered into a truce in 1618, allowing Poland to start preparing to attack the Turks. In earlier decades, French interests had brought Henri, duc d'Anjou, to the throne of Poland and Lithuania before he returned to France to become King Henri III. Lorraine also had interests there: one of the founders of the ruling house of Lorraine was Christina of Denmark, whose connections throughout the Baltic, North Sea, France, Holy Roman Empire, and Italy no doubt continued to associate the interests of Lorraine with the other rulers. If Descartes were seeking attachment to a person or cause, or were seeking information about current events for anyone in France or Lorraine—or for anyone else—there would have been many opportunities to gather news.

Baillet reports that Descartes began and ended his Baltic visit at Stettin/Szczecin, then under the suzerainty of Pomerania, allied with Sweden but coveted by Brandenburg.[23] In Stettin, discussions were under way about the duchy of Prussia, which had been taken over by the Hohenzollern elector of Brandenburg, George William, but which remained a fief of the king of Poland. But he also visited several other port cities. Descartes traveled back and forth in the region for several weeks.

Perhaps he was on business for Brandenburg's adversary, the Count Palatine-Neuburg. Brandenburg was at war with Palatinate-Neuburg in the Rhineland: Brandenburg possessed Cleves, and Neuburg held Jülich. Baillet's denial that Descartes had fought with the Ottomans might suggest that he was working against Polish-Brandenburg interests. If he were conveying any messages, he would have found a number of sovereign princes nearby. But whatever his reasons, after some time Descartes again decided to go on, this time traveling west to Mecklenburg and Holstein.

In other words, for some reason as the year 1621 wound down, Descartes was moving about along the Germanic edge of the Baltic. From Pomerania westward the territories were part of the Lower Saxon Circle, whose most senior figure was the king of Denmark. When not stopping at the free cities of the former Hanseatic League, Descartes moved through Danish lands. Christian IV yet remained officially neutral in the wars, although sending financial support to Brandenburg and to Frederick of the Palatinate (recent king of Bohemia), who was the husband of his niece, the Winter Queen Elizabeth Stuart. In returning to places he may have already visited in the summer of 1619, was Descartes assessing Danish intentions, or had he even been in their service since his first visit to Copenhagen? Or was he simply passing through because the roads west remained quiet? When King Christian finally entered the war against the emperor in 1625, he would be supported in part by French subsidies.

In November 1621, Descartes started on a final leg home via Holland. The account of Baillet suggests the ending of an enterprise of some sort. Descartes sold his horses and most of his baggage, and he released all his retainers except his valet. One might guess that he did so at the great commercial city of Lübeck, for he then made a short trip overland and at either Hamburg or Glückstadt (on the Elbe) boarded a ship for East Frisia, intending to go on from Emden to West Frisia. From there he could easily travel to Amsterdam and other places familiar from his recent stay.[24]

But again he was in mortal danger. Having sold most of his things and paid off most of his entourage, he must now have been traveling almost alone and probably not only with his remaining baggage but with money in hand. When he rented a small ship and found some "volunteer" sailors to take it from Emden to West Friesland—presumably paying to accompany them on a trip they were already making—the crew thought he was a foreign merchant. Because Descartes spoke only to his valet, and in French, they did not realize that he understood their language (meaning that arrangements must have been made without Descartes speaking with them). The sailors plotted to murder him as soon as he fell asleep and to toss him overboard, taking all his money and goods. But Descartes overheard and understood. Biding his time, he suddenly jumped up fiercely, brandished his sword, and explained to them in their own tongue, in no uncertain terms, that he would cut them to shreds if they did not proceed as originally intended. They could not stand up to the strong spirit of the chevalier, Baillet says.[25] He had been in the wars, and no doubt well understood how to assume the grim face of being in deadly earnest.

Arriving safely in Holland, about four months after the disaster at Neuhäsel, Descartes at last paused for the winter of 1621–22. But even now he seems not to have been acting simply on his own account, for he oddly did not visit his former friend and mentor, Isaac Beeckman. Back in France, in the middle of December, the king's former favorite, the duc de Luynes, had died, giving free field for the rise of Richelieu, who would be given his cardinal's hat in April. If Descartes was associated with the government, his managers might have been assessing the direction policy would take.

But Descartes remained active. Baillet reports that he visited the seat of the government of the Dutch Republic in The Hague on three separate occasions, once to see the States General, once to see the court of the Prince of Orange, and once to see the unfortunate Winter Queen, Elizabeth. Was he simply a tourist? He remained keenly interested in the current military situation. The Twelve Years Truce between the Dutch

Republic and Spain had recently ended (in April 1621), and the Habsburg general, Spinola, had quickly organized a siege to take Jülich with the support of Wolfgang Wilhelm; the city was held by Dutch troops on behalf of the Hohenzollerns of Brandenburg, who claimed neighboring Cleves but which was also under Dutch control (and where the house of Lorraine also had interests).[26] Baillet says that Descartes went to watch the progress of the siege and stayed until it was over. By implication, he continued to be free to move about among the imperial forces, again suggesting an association with Count Wolfgang Wilhelm or Lorraine. Spinola successfully took the city in February, after a five-month struggle. Descartes then went on to the court in Brussels, no doubt to join in the formal celebration of the victory but possibly to keep an eye on what was being planned next in the wars for the Rhineland.

Again in France

Finally Descartes returned to France, first to Rouen and then to his father's house at Rennes, which, according to Baillet, he reached in the middle of March. And indeed, there is archival evidence of his signing a document in Rennes on April 3, 1622, giving his brother power of attorney to sell the lands he had inherited from his mother.[27] This document authorized the completion of the business that his brother had begun in 1618–19, and it may have transferred a sum of money into his hands. Perhaps he had debts to pay. But Descartes would not stay long.

The France to which he returned was in the throes of a new round of internal warfare. Louis XIII was pushing aside the *politique* policies of Henri IV and Marie de Medici in order to destroy the Huguenots. The ruling family had also, for the moment, made peace with itself, allowing it to present its ambitions less as a personal struggle to retain power than as an attempt to unify the kingdom. Among the other nobles of the day, Louis XIII commanded the greatest ability to raise military forces. But the duc de Montmorency—who had protected Théophile and would soon side with Gaston—warned that the attempt to ruin the Huguenots was an attempt to bring the other rival houses to heel, for it was the Protestants who subsidized the princes and *les grands*.[28]

Descartes's friend Balzac had been in the midst of it. At the time Descartes had left France following the assassination of Concino Concini, Marie de Medici fled to the royal château at Blois, where she was told to remain. But one night in late February 1619, she made her escape and joined an armed escort of the duc d'Épernon and her ten-year-old son, Gaston d'Orléans, and together they headed an aristocratic revolt against her other son, the king. They occupied the seigneurie de Balzac. In the previous year, Balzac had published, anonymously, an open letter on behalf of his patron d'Épernon against the Keeper of the Seals, Guillaume Du Vair, and from then on he wrote often on the duc's behalf against his enemies. D'Épernon thought so well of Balzac's talents that he proposed him as personal secretary to the queen mother. One of her former advisers, Armand Jean du Plessis, soon to be known as Cardinal Richelieu, was also interviewed, in his case for the position of confessor, and he insisted on negotiations with Louis. D'Épernon decided to support a reconciliation, and in the home of Balzac's father, d'Épernon and the queen mother received a delegation of high-ranking negotiators from the king. Balzac put his ability to use by writing three letters to Louis XIII justifying the duc's actions as necessary for the good of the kingdom. Marie and Louis signed the Treaty of Angoulême on August 10, 1619. But with Richelieu on the scene, Balzac decided not to serve the queen mother but the youngest son of d'Épernon, Louis de La Valette, archbishop of Toulouse, accompanying him to Rome at the end of April.[29] (Balzac returned from Rome in April 1622[30] but then stayed away from court, so on Descartes's return to France, Balzac would not have been in any position to help a former acquaintance in person.)

In 1622, the royal family having been reunited, Louis XIII began a new round of religious wars, accompanied by a period of severe cultural reaction. Balzac had gone to Rome in the service of the French bishop in whose archdiocese the city government had just horribly executed Giulio Cesare Vanini for atheism and blasphemy. The king him-

Map 2. Places associated with Descartes in northwestern Europe. Drawn by Lyse Messmer.

self was pious. Those around Marie de Medici held a variety of views but were generally committed Counter-Reformation Catholics: d'Épernon himself, for example, had recently been warring on Huguenots in Guyenne, and Richelieu had publicly advocated France's alignment with the Council of Trent. Queen Anne of Austria, too, was devout, and she also secretly kept in close touch with her father in Spain. The policies of the ruling family and its supporters and retainers were consequently increasingly distasteful both to *politiques*, who put loyalty to the kingdom above religion, and to Huguenot barons, who took steps to organize themselves. From the spring of 1621 onward, therefore, the king had been moving through the country with the queen and queen mother, his favorite the Duke de Luynes, the Prince de Condé, and other nobles, officers, and soldiers, restoring his authority by forcibly seizing disobedient towns, which were often led by Huguenots, before spending the winter of 1621–22 in Paris. (In 1621 the queen mother was restored to the king's council and began to refurbish Luxembourg Palace with the help of the Flemish painter Peter Paul Rubens; in December, Luynes died of a fever, leaving his recent wife, Marie de Rohan, widowed.)

Then, about the time that Descartes reentered France, the king set off on campaign again. Louis XIII's army reached Nantes later in the spring of 1622, and headed into Poitou, sweeping up the forces of the Huguenot duc du Soubise along the coast. By early summer the king's troops were in the south, continuing to take Huguenot cities, sometimes committing horrible atrocities, as at Nègrepelisse. But with the Duke of Rohan keeping up the fight against the king in the south, and with a Protestant German army under Count Mansfeld entering the French north, in Champaign, to create a diversion, a long siege at Montpellier devolved into negotiations and a treaty in October, for the moment reaffirming the cohabitation of the two religious parties in the kingdom.[31]

Descartes's movements in this period are little known but intriguing. Baillet explains that having inherited his portion of his mother's estates, which were in Poitou, he decided to look them over, and then spent the summer in Châtelleraut and Poitiers while his father went to Chavagnes, in the diocese of Nantes, where his second wife had property.[32] Father and son had hardly ever lived together, so it is no surprise that they did not stick together now. More curiously, however, they were both moving in the wake of the king's forces.

Although the younger Descartes now had plenty of military and diplomatic experience, Baillet does not mention him joining in, but one wonders. One of the great lords whose patronage he may have sought, Charles de Guise, led the action against Soubise in the region of La Rochelle that led to a royalist victory in October. Other great lords with whom he had previous contact, the duc d'Angoulême and Béthune, the comte de Selles, had returned from their embassy to the empire and were now with the king. The Balzac family patron, d'Épernon, was there, too. So was one of Marie's favorites, Bassompierre, being Vanini's former protector and well connected to the Lorrainers: he had been on an embassy to Spain in 1621 but now fought against the Huguenots so well that in 1622 Louis created him a marshal of France. If the martially experienced Descartes was seeking patronage, he may well have been near the court, on the move.

The sense that Descartes must have been looking for a place is shared by Baillet, who mentions that Descartes's father had nothing to offer him. Descartes therefore decided to return to Paris early in 1623, Baillet explaining that at last the plague had left it.[33] But that reason does not make sense: while plague was certainly an endemic problem in the period, there is no evidence of a major outbreak in Paris prior to 1623. Instead, it began to become troublesome later that same year.[34] If Baillet is correct about when he visited Paris, Descartes arrived there at the end of February 1623 (at the beginning of Lent), a full year after his own return to France but just six weeks after the return of the king and his court to the capital city. There Descartes "tasted the peace" that Louis had brought his people by the suppression of the rebels, which suggests he was continuing to associate himself with the monarchy.[35] Where had he spent the summer, autumn, and early winter? Given his Italian travels, which come next, a good guess is that the sieur du Perron was in the service of Béthune, d'Angoulême, or another royal official, traveling with the court on campaign.

Paris and the Rosicrucian Scare

When Descartes returned to Paris, Baillet wrote, his friends were happy to learn from him some of the inside news about events that had transpired in the empire. Perhaps Descartes kept up with contacts connected to imperial interests. The war in the empire had continued to go in favor of the duke of Bavaria. A conference had been convened in Brussels not long after Descartes left there, where England hoped to negotiate a peace, but Maximilian of Bavaria continued to act aggressively, seizing Heidelberg and Mannheim. In February 1623—about the time Descartes reentered Paris—Emperor Ferdinand rewarded Maximilian for his efforts by transferring to him the territories of Frederick and his Palatine electoral dignity, giving a secure majority of imperial electoral votes to the Catholics. Even Spain thought this was a step too far in governing the empire, however, and these actions began to turn French policy, too, toward confrontation with Ferdinand. In the aftermath, Lorraine would be caught between France and the Holy Roman Empire.

But the immediate problem for France now lay in the southern Alps. The news from Regensburg about Maximilian coincided with Descartes's reappearance in Paris, but he was soon on the move once again. According to Baillet, he left Paris after only two months, again moving quickly. He returned to Brittany in May (to consult with his elder brother and possibly his father) and then to Poitou in June and July to see to the sale of his estates—while keeping his seigneurial title—before paying another brief visit to Paris on his way to Italy. Among those he revisited may have been Claude Mydorge and his colleague Claude Hardy.[36] He probably also met for the first time Marin Mersenne, who in 1619 had taken up residence in Paris as corector of the Minims based at l'Annociade, their house just to the north of the new Place Royale. (If Descartes was attending the court, he might also have noticed two incognito English gentlemen passing through: the Duke of Buckingham and the Prince of Wales, who were on their way to Madrid.)

But his intention to leave again so soon puzzled his friends. His explanation might have been dissimulating: "That a journey beyond the *Alps*, would be much to his advantage for the instructing him in business, and gain some experience in the World, and get acquaintance with men verst in Worldly affairs, which he had not yet done, adding, that, tho *he might not return Richer, yet at least he would come back from thence more capable for business*" (emphasis in the original).[37] Or perhaps he really did wish to join in the world of commercial exchange, to which he had been much exposed in The Netherlands and the Baltic. He would certainly have known that during a period of military buildup there was money to be made. Money was beginning to rule the world, he had acquired more than the practical mathematics necessary for keeping the books, he had demonstrated an ability to run risks, and he was disposing of his estates so as to obtain money from their sale, so his explanation is not as ridiculous as it may sound. In any case, he must have decided to prepare the grounds for this new move in late 1622 or early 1623, since the sales could not be accomplished overnight.

Baillet is very clear that Descartes had by now given up most of his former interests in mathematics and natural philosophy. He quotes Descartes writing in 1638 that fifteen years earlier (that is, 1623) he had "quite laid aside Geometry, and would never more meddle with the solution of any Probleme, but only at the request of some friend."[38] Descartes had also decided that mathematics was of little importance when studied for its own sake instead of for use. He forgot most of his arithmetic, although his love for geometry persisted a while longer. But he remained interested in reviving the study of algebra, and in what he called "mathesis," or universal mathematics. According to Baillet, Descartes also turned to the study of "physic" (*physique*), or medicine, which in determining the things that made for a good life also led to ethics. Although Descartes had not found the secret for preserving life, he had found out how not to fear death.[39] John Schuster's recent study of the young Descartes's intellectual development gives no evidence for any interest in philosophy more generally at the time.[40]

Before leaving Paris, however, Descartes became implicated in the public fear of subversion by the Rosicrucians. In 1623, printed and manuscript handbills and broadsides began to appear in the city, proclaiming that leaders of the brotherhood had come to Paris incognito, possessing wonderful abilities to help their fellow men, and inviting the public to a meeting. A recent study of the episode argues that it was a hoax perpetrated by a medical student, Étienne Chaume, and his friends.[41] Some French authors had already heard about the Rosicrucian brotherhood, while versions of the hermetic philosophy that underpinned the Rosicrucian tracts were widespread in the writings of Raymond Lull, cabalism, astrology, and alchemy, to say nothing of the continued interest in the prophecies of Nostradamus and other occult studies. Some French commentators at the time (such as Baillet in later years) also associated the Rosicrucians with a heretical sect known as the "Alumbrados" that had just been identified in Seville, where they were eliminated by the Inquisition.[42] Baillet indicates that Descartes, having just come from Germany, was accused of being one of the six leaders of the Rosicrucians who had secretly come to take over France, and that Mersenne in particular—who was just then bringing out major books against all kinds of unorthodox views of God and nature—asked him bluntly about his possible involvement. The answer Descartes gave, according to Baillet, was that the brethren were reputed to be invisible, but that he himself was not, and so he continued to show himself to a "great Concourse of People" and the rumor died.[43]

But this was no simple jest. Whereas Baillet places the episode early after Descartes's return to Paris (that is, in March or April), Didier Kahn's investigations show the handbills to have appeared between mid-June and the end of July, meaning that the public association between Descartes and the Rosicrucians would likely have been made during Descartes's brief visit to the city in August.[44] That was a particularly dangerous moment. On July 11, 1623, the Parlement of Paris had issued orders for the arrest of Balzac's former friend, Théophile, whom the Jesuit father Garasse accused of being "the head of the band of atheists" (*le chef de la bande athéiste*) now that Vanini had been eliminated. In August, Théophile was sentenced in absentia to make a public apology (*amende honorable*) before being burned at the stake; instead, he went into hiding, soon to be arrested at the border as he tried to make his escape to England. Influential friends would make sure that he was only held in prison without trial, and when he finally came to court in 1625 he was acquitted, given that the Jesuit intrigues behind the affair were so obvious. But his example certainly discomfited anyone of libertine sympathies.[45]

Moreover, among the charges against Théophile was that he had been one of the leaders of the Rosicrucians because a Rosicrucian manuscript had been found among his papers. Théophile might be a scandalous poet, but a long tradition also associated mystical philosophy with eros.[46] There had also been other early associations between libertines and Rosicrucian works. For instance, a historian of the movement, John Montgomery, noted that the Rosicrucian *General Reformation* (*Algemeine und General Reformation der gantzen weiten Welt*, 1614) was closely based on a work of a satirist and architect from Loreto, Trajano Boccalini, who died in Venice in 1613, perhaps beaten to death for his freewheeling views.[47] Théophile explained that the manuscript found among his possessions had been planted, and even went so far as to say that the Rosicrucian placards had been conceived as a means to lure freethinkers into showing themselves for the purposes of identification and suppression.[48]

He may have had a point. Recent work has pointed to possible connections between the Rosicrucian publications and the Jesuits. A few years later a conspiracy theory originating with them had it that in 1621 an important group of influential intellectuals was plotting to replace Catholicism with Deism. The plot supposedly involved people later identified with Jansenism, such as Duvergier de Hauranne, better known as Saint-Cyran, and possibly even Antoine Arnauld (then a boy). Anti-Jesuit writers including Blaise Pascal and Pierre Bayle would attack this purported conspiracy as a dark and dangerous fantasy meant to create harm.[49] There is even a real possibility that some of the earliest reports of the Rosicrucian works come from Adam Haslmayr, a schoolmaster, musician, and alchemist from the Tyrol, a subject of Archduke Maximilian of Austria; Maximilian is sometimes described as "completely devoted to the Jesuits and the Inquisition." If the Jesuits furthered the publication of the Rosicrucian tracts to root out heterodoxy, they had some success: a friend of Haslmayr's was implicated in the publishing and imprisoned; the supposed Rosicrucian Haslmayr was himself sent to the galleys of Genoa for four and a half years.[50] But whatever the merits of Théophile's suspicion about supposed Rosicrucianism as a stalking horse to identify and condemn libertines, such people as Garasse were not only clearly drawing associations between Rosicrucianism, epicureanism, and atheism but also working the judicial system to have Vanini executed and Théophile set on a path intended to end at the same place.[51] The mood must have been foul.

For Descartes to be called out by Mersenne as a Rosicrucian at that moment was, then, no joke. The witty answer that Baillet says he gave suggests that Descartes remained faithful to the chivalric code and gave no sign of fear. But he was in any case heading for Italy. By the end of August he was gone.

The Valtellina and Rome

Descartes again traveled to a place at the heart of current affairs: to the Alpine valleys known as the Valtellina. The French would postpone military action there for some months, but conflict in the Italian Alps was on the boil. Baillet says that Descartes's move was prompted by the news in March of the death of Monsieur Sain, who had been a tax collector (*controlleur des tailles*) for Châtelleraut. Descartes was related on his mother's side to Sain, who was also the husband of his godmother. Actually, in the Estates-General of 1614, Sain was listed as from Tours, councilor of the king and treasurer general of France.[52] Baillet believed that Sain had also taken on the position of commissary general (*commissaire general des vivres*) for the army in the Piedmont—such positions, looking after the provisioning of an army, could be extremely lucrative.[53] The reported "pretext" (*prétexte*) for Descartes's journey, then, which he gave to his friends, was that he was both looking into the affairs of his relative and seeing if he himself could obtain the post of intendant of the army. Just in those years such offices were becoming the norm in the French forces, with the intendant looking after the justice and discipline, and finances, of the organization on behalf of the crown. (Intendants also oversaw the provision of military hospitals and care of the wounded, perhaps giving Descartes an additional nudge toward an interest in medicine.) Some of those who held such offices rose to high places in the royal administration.[54] Descartes seems to have been seeking to make visible in France some of the skills he had acquired abroad and by doing so to gain favor, since he told his friends that he would learn how to accomplish things in the real world.[55] In retrospect, his explanation does seem a pretext, since he did not gain the office, but he seems to have returned with plenty of cash in hand.

He headed straight to the center of action, the Valtellina. A few months earlier those valleys were on the minds of everyone concerned with international affairs, since the so-called Spanish Road ran through them. If Habsburg troops and supplies were to cross from Spanish-held Milan northeast into Austria and Germany, or northwest into Lorraine and the low countries, they had to pass through the Valtellina. The chief overlords of the valleys, the Protestant Grisons, had been supporters of Frederick's election as king of Bohemia, but a large portion of the population was Catholic. Hard-line Catholic incendiaries had been able to stir up the people of the valleys, who appealed to Spain for protection, and troops from Milan moved in, beginning a series of bloody conflicts that also came to involve the Emperor Ferdinand's brother, Archduke Leopold of Austria. The marquis de Bassompierre had been sent to Spain to insist that the valleys remain open and that the Protestants remain free to practice their religion, points that were inscribed in the Treaty of Madrid of April 1621.

But by building a series of fortifications to protect their position in the valleys, the Spaniards now not only secured the way north but also divided Savoy from Venice.[56] France joined Venice and Savoy in an alliance to force Spain and Austria out of the Valtellina (the Treaty of Lyons, of February 1623). The alliance demanded the withdrawal of Habsburg forces, and Spain backed down to the extent of agreeing to turn its forts over to neutrals, the troops of the papacy. Even more important, following the death of the pope in July the non-Spanish Catholic allies managed to negotiate the election of a reforming and Francophile prince of Rome, Maffeo Barberini. Open war in the Alps would be postponed until the end of 1624, but the diplomatic confrontations were ratcheting up.

It was just then, in August 1623, when Descartes headed straight for the Valtellina. Baillet says that he traveled to Basel and then Zurich, and from there he could have moved farther eastward to arrive at his destination not long after. He would easily have been in the Valtellina long before winter. Baillet thinks that in either Chiavenna or Tirano he must have met the marquis de Bagni, the cardinal and papal envoy who was overseeing the handover of fortifications from Spanish to papal troops (not to be confused with Jean François Guidi, the papal nuncio who was sometimes also known as Bagni and later also befriended Descartes).[57] Nearby were French forces under the command of François-Annibal d'Estrées, marquis de Coeuvres (brother of Henri IV's favorite mistress, Gabrielle d'Estrées), who would lead the French attack on the Valtellina at the end of November 1624. With him was Jean-Jacques de Haraucourt, sieur de Haraucourt, from Lorraine, a favorite of Charles de Lorraine, duc de Guise,[58] with whom Descartes seems to have established connections a few years earlier. It is a good guess that Descartes considered himself to be a competitive candidate for intendant, using his personal connections to gain a hearing and offering himself in place of a relative who had already been engaged in supplying the forces.

But he seems to have found service not in administration but in diplomacy—or perhaps the intendancy had been a pretext after all. With a new pope on the throne of Saint Peter, negotiations about Roman troops taking over from the Spaniards were ongoing, and so, Baillet says, Descartes went on to the Tyrol and Venice, stopping on the way at Innsbruck to visit the court of the Archduke Leopold, who also had troops in the Valtellina. He seems to have spent the winter in the mountains, but he arrived in Venice around Lent in 1624. He made sure to be there by Rogation Day—or Ascension Day, the fortieth day of Easter—to see the annual Sensa, or Wedding of the Sea, performed by the doge (which came early in May that year).[59] After he completed the business that called him to Venice, he also fulfilled his former promise by making a pilgrimage to the Virgin Mary of Loreto. If he was finished with his business in Venice by mid-May, and made the roughly 250-mile (400-kilometer) pilgrimage on foot, as he said he would, he might have been able to return to the world by the end of June at the latest. Did he return to the Valtellina or otherwise continue to stay involved with current events?

But then, Baillet writes, he thought again of his "pretext" of trying to obtain the intendancy of the French army in the Piedmont, traveling to Rome to negotiate about it. Curiously, however, his visit to Venice had coincided with an announcement by the new pope, Urban VIII, that there would be a jubilee year in Rome beginning at Christmas 1624. Coincidentally, perhaps, Descartes arrived in Rome just before the jubilee opened, at Advent, the end of November; it also happened to be the moment when d'Estrées launched his assault in the Valtellina. In Rome, Descartes saw Władysław IV Vasa of Poland, who had been fighting the Ottomans in Moldavia during Descartes's passage through that region and had been at part of the negotiations between Brandenburg and Prussia; he again saw Archduke Leopold, whom he had recently visited at Innsbruck; and he met the comte de Chiavenne, from where he had started his visit to the Alps. Baillet remarks that because of the jubilee, Descartes encountered people from every part of Europe, and given his passion for investigating human nature, he spent far more of his time looking into the affairs of men than paying attention to either the ancient or modern sights of Rome. He stayed until spring 1625, when he accompanied the pope's nephew, Cardinal Francesco Barberini, on the first stages of his embassy to Paris.[60]

Not much of this makes sense if we think of Descartes as an independent actor who was simply traveling for his own interest and edification. Why would he first be traveling in a mountainous region in the autumn and winter while war was impending? It is an odd place to begin a touristic visit to Venice. His pilgrimage to Loreto was added to fulfill a promise he had made to himself. But then why go to Rome to negotiate about a French intendancy in the Alps? Why remain in the Eternal City for several months and not have a look at the antiquities and the impressive new buildings? He was too old to have a tutor in tow, and his behavior was quite unlike a grand tour as made by other young gentlemen.[61] Descartes's later biographer, Charles Adam, therefore considered Baillet to be filling in some unknown years with speculation about the travels based on Montaigne's literary account of his own visit to Italy (which was to find a cure at the mineral baths for his illnesses—but Descartes was healthy). A recent biographer, Stephen Gaukroger, agrees that Baillet's report must have been modeled on Montaigne's. Another, Geneviève Rodis-Lewis, is suspicious of the whole Italian journey, moving the narrative back to his return to France in May 1625 as quickly as possible. The expert editors of the *Historical Dictionary of Descartes and Cartesian Philosophy* are also doubtful, simply noting that the trip to Italy was required by "other financial matters" and skipping over any speculations about what he did there. If Descartes was simply acting without purpose, it would all be most mysterious.[62]

But between June and November 1624 Descartes would have had time to return to the French army in the Piedmont to negotiate about a post, or to do countless other things. Then, when he went to Rome, Baillet reminds us, the French ambassador was Philippe de Béthune, whom we know Descartes had encountered before. For his services to the crown Béthune had been made a *Chevalier de l'ordre du Saint-Esprit* (something like a Knight of the Bath in England).[63] He had been named to the post of special envoy (*ambassade extraordinaire*) to the Holy See in April 1624, about the time that Descartes was in Venice.[64] Béthune was on good personal terms with the new pope: he had spent many previous years in Rome as a French ambassador, and he had helped ease the way for the former Cardinal Barberini when he traveled to Paris in 1601 as the papal legate to Henri IV; both men shared a Jesuit education; both patronized Caravaggio and other painters; and both shared interests in literature and natural philosophy. (In the comte's case, his collections of "naturalistic" art were extremely rich, later attracting the envy of Queen Christina of Sweden.)

Béthune had been sent to Rome to try to sort out the business in the Valtellina. When d'Estrées finally made his move in November 1624, he quickly took the valleys, as the forces of the pope offered little resistance except at Riva and Chiavenna. Spain in turn allied with Tuscany, Parma and Modena, Genoa, and Lucca; a French-Savoyard army subsequently began an assault on the republic of Genoa in February 1625 while Venice promised to move against Milan. As friendly as the new pope might be toward France, he could not accede to these actions, and he insisted on the return of the fortresses. Descartes had been circulating among the diplomatic hot spots and then came to Rome just when the French attacked, when both Béthune and the papacy needed knowledgeable people nearby on a daily basis. With hostilities now under way, however, the original purpose of Béthune's embassy was coming to an end. In the spring, the pope went around Béthune by sending his nephew to Paris for negotiations. Descartes set out with that group. How could he not have been connected to events, as before?

The Lorraine connection might have still been active, too. At the beginning of 1625, someone working for Richelieu's confidant, Father Joseph, transmitted a plan to Louis XIII from Archduke Leopold that would have made for a "Holy Alliance": aside from France, the archduke of Tuscany would control southern Europe, the duke of Bavaria would rule the Holy Roman Empire, and the duke of Lorraine would rule the low countries.[65] Strange dreams were continuing to surface.

Nevertheless, being in Rome at that moment would have a profound effect for Descartes's future intellectual program. With the new pope in place, the city was abuzz with intellectual promise. Intellectual reformers anticipated that they would be allowed to set Catholic thinking back on the surest path toward a universal understanding of God, through the study of his creation. Among those making a stir in print as well as conversation were the "new philosophers," who often called themselves "virtuosi." Libertines such as Giulio Cesare Vanini had excited court circles in France about the view from Padua more than a decade earlier. In Padua, freethinking philosophers and physicians continued to develop ways to show that mathematical and physical investigations of the material world, on the basis of epicurean atomism, could explain all phenomena. Cesare Cremonini was teaching such views at the time Descartes visited Venice. Additionally, Santorio Santorio had just stepped down from the Paduan chair of theoretical medicine, having made the case for explanations of living bodies based on number and matter, with the first edition of his *Ars de statica medica* (1614) presenting one of the most important early works to use quantitative approaches to medicine.[66] By several times a day systematically weighing himself, his food and drink, and all his evacuations, he proved the existence of "insensible perspiration." Whether Descartes was directly connected with Santorio is unknown, but if he did express an interest in medical theory while he was in Italy, it is hard to imagine that Santorio's work would not have been a part of the conversation. In Rome itself, the Calabrian monk, Tommaso Campanella, objected to the Paduan line as atheistical but at the same time turned anti-Aristotelianism into a paean for the unity of corporeal nature and animal *spiritus*. He was then helping Urban VIII to combat astrological rumors of the pope's impending death.[67]

Most notably, a friend of the new pope, and of Cremonini and Santorio, Galileo Galilei—former mathematician and engineer of Padua, now philosopher to the grand duke of Florence—had recently published *The Assayer*. A copy of the book had been presented to Urban VIII with his approbation on October 27, 1623, about the time that Descartes arrived in the Valtellina, and everyone was discussing it.[68] Taking the form of a witty and entertaining discourse, it offered a program not only for defending heliocentrism but also for explaining all the knowledge we have of the world as coming via the senses, while they in turn were explained as responding to particles of matter. Heat, for instance, was not a quality per se but a sensation resulting from fiery particles. The material stuff of which the world was made could be subjected to analytical investigation and described mathematically according to its movement in the three dimensions.[69] Even a year later, the book must have been among the many subjects discussed by the French delegation. For Descartes, with a sound familiarity with the world from which Galileo had emerged, hearing the familiar echoes of Epicureanism reformulated in terms that a military engineer could appreciate must have been exciting. In all likelihood, like the meeting with Beeckman in 1618, it pulled Descartes back into the orbit of natural philosophical debates, for he was clearly reengaged in them by the time he returned to Paris.

Perhaps Descartes even met Galileo. There was more than enough time for a journey to Florence between his visit to Loreto and his entry into Rome (which was perhaps as long as five months); from April through June 1624 Galileo was himself in Rome, but Descartes could have visited Florence later in the summer or autumn. Having returned from his recent successes in Rome, the sixty-year-old Galileo was then living south of the Arno in the hills overlooking Florence and tending his garden.[70] But he was not a recluse, and he received many visitors. If Descartes's interest in mathematics or philosophy had been rekindled—unknown but likely—then he might well have taken the time to visit. Or perhaps he stopped by when he departed Rome with Francesco Barberini, who had recently been tutored by, and would remain personally very supportive of, the virtuoso; also accompanying Barberini was Giovanni di Guevara, who had been appointed to examine *The Assayer* with a critical eye but instead became a defender of it.[71] Moreover, Descartes's first biographer, Pierre Borel, insisted that the two had met.[72] In 1633, however, when Galileo was famously condemned by the Inquisition, Descartes wrote Marin Mersenne a series of letters disavowing any interest in Galileo's work while at the same time saying that he was burying all that he himself had been writing for the past three years because he did not wish to contradict any official declarations of the church.[73] When asked directly by Mersenne, he said he never had seen Galileo (*je ne l' ay jamais vû*), nor communicated with him, nor found anything in his books that gave rise to a feeling of envy or admiration other than his views on music. That is a full and complete denial, which Baillet accepted as a declaration of the truth.[74] But it is clear that Descartes knew and respected Galileo's work, if no more.

Human interventions continued to demonstrate other ways in which power over bodies might affect minds, however. Cremonini himself very carefully avoided confrontation with the Inquisition, even refusing to look into his friend Galileo's telescope. Despite Campanella's current assistance to the pope, most of his works were written in prison, where after initial severe torture he spent almost twenty-seven years. (He was released to Rome in 1626, escaping to France in 1634.) More immediately, Descartes arrived in Rome only a few weeks before the burning of the remains of the ecumenical Cardinal Marc Antonio de Dominis and his books in the Campo dei Fiori of Rome on December 21, 1624. De Dominis had once been an archbishop who came to resent papal interference and sided with Venice during the papal interdict of 1606–7. In Venice he also wrote a work on how the refractions and reflections of light on droplets of water could create the effects of the rainbow, a theory often credited to Descartes.[75] Afterward threatened by the Inquisition, de Dominis left for England, where he wrote about the superiority of the bishops over papal monarchy and saw Sarpi's *History of the Council of Trent* through the press. But unhappy in London, he went to Brussels, recanted, returned to Rome, and was readmitted to offices in the church. For confessing that he believed that a reunion of the Christian churches was possible, however, he was imprisoned, and was awaiting trial by the Inquisition when he died in September. After the posthumous verdict against him, his bodily remains were dragged through the streets and publicly burned. Being in the city at the time, Descartes could not have missed the message being sent about the political pretensions of Rome, even with a cultivated Barberini in charge.

At the end of March 1625, Descartes set off to Paris in the company of Cardinal Francesco Barberini's delegation, which would make demands related to the Valtellina. Baillet says that he felt it was an important courtesy to help in this way, indicating a friendly relationship with the cardinal.[76]

But on the way, Descartes left Barberini to attend the siege of Gavi. France had supported the attack on Genoa by the Duke of Savoy, contributing French troops under François de Bonne, duc de Lesdiguières and constable of France, reinforced by soldiers from the Valtellina under d'Estrées. The allied army had moved quickly through Genoese territory but then stopped to take the fortress at Gavi, which resisted. It capitulated on April 22, but the delay had given time to the city of Genoa to prepare its own defense, and with naval support from the Dutch and British failing to materialize, the Spaniards broke through the French naval blockade in August to relieve the city.[77] Descartes followed the French troops for a while after Gavi, but then he turned toward Turin, where he met Christine, the young princess of Piedmont (and daughter of Marie de Medici): her husband, Victor Amadeus, was with his father the Duke of Savoy outside Genoa. Perhaps Descartes was continuing to seek patronage in service to one of the armies, but it seems more likely that he was acting on behalf of someone.

At last Descartes returned to France via the Swiss passes—where he took notice of avalanches—meaning that he must have gone through Aosta and Chamonix to Geneva and then Lyon.[78] From Lyon and Poitou he sent a now-lost letter to his godmother explaining what he had been able to discover about her husband's business with the army of the Piedmont, and another from Châtelleraut to his father dated June 24, 1625, discussing the possibility of taking on the position of lieutenant general in the city—which, as we have seen, coincided with aristocratic plotting and Richelieu's cold-blooded reaction; it would not work out.[79] After the better part of a decade, Descartes had returned to France. But he would find no peace.

*

Descartes was now well experienced in the ways of the world. As Baillet insists, his time had been spent far more on the study of humankind than on anything else. His movements from 1621 to 1625 were closely associated with the major events in Europe, first with the imperial armies in Germany, but also with the French embassy sent to forestall the war. For a while he put his engineering abilities to use, but the death of the count of Bucquoy meant the death of a possible military patron. After the defeats in Hungary, he had moved through the southern Baltic, perhaps on business related to the junior Palatine house. When back in the low countries, he certainly kept a watch on events in Jülich from the imperial side. Then in France he is likely to have been with the king's army on campaign, too. He stopped only briefly in Paris, coincidently with the Rosicrucian scare, since the sale of his estates went forward and he made a rapid departure into the Italian Alps, in the midst of war, perhaps seeking a military administrative office but certainly moving among the diplomatic hot spots. Then, after fulfilling his vow to visit Loreto, he was in Rome for the opening of the papal jubilee, just when the French attack on the Valtellina began, again visiting with dignitaries and keeping a watch on the human comedy rather than attending to the sights. He must have encountered the latest discussions about the new philosophy while there, as well. His return to France began in the company of the papal envoy but was delayed by yet further military interests. The certainties of his world had come apart, he had survived real threats of death on more than one occasion, and he was truly speaking to power. He kept moving.

How one wishes to know for whom he was working or from whom he was seeking office, or to know the several persons! Bourbon, Guise, Lorraine, Sully, Wittelsbach, perhaps Denmark, Barberini, even Savoy, each in their turn or all together might apply, but certainly not either Jesuits or Huguenots. The simplest explanation holding them together is that Descartes remained loyal to the interests of the queen mother. In any case, none of the possible connections would have identified him as disloyal to France, although from the mid-1620s onward, that would change. In a different world some decades later, Baillet clearly thought it best to avoid the subject. The efforts of Louis XIII to eliminate all internal opposition and to physically control Savoy and Lorraine would continue to stir many nobles to dissent, sometimes to take up arms against him. Many of the persons in the kingdom whom Descartes is likely to have served would, in the end, not sit happily at the young king's table. By Baillet's day, Lorraine, for example, had been reclaimed by France, driving the ruling house into service to Louis XIV's enemies, the Viennese Habsburgs.

The sieur du Perron may, however, have stumbled across another approach to bridging division and uniting the world: he had bumped up against the illuminating excitement of Galileo's physicalist explanations of nature and our body's sensations of the world. Resolving human conflict not by burning heretics and atheists, or executing dissenters, but by burning away all the fantasies that could not be established on the basis of clear and distinct ideas, the kind of evidence that existed in three dimensions rather than in unverifiable qualitative attributes: creating harmony and bodily betterment by establishing the real facts: a noble, universal, millenarian dream. We know at least that he placed much weight on dreams and the truths spoken by poets. Perhaps powers other than the sword now pointed a way forward.

PART 5

The Struggle for France

From June 1625 for the next three and a half years, René Descartes would base his activities in Paris. But they are just as mysterious as his earlier years abroad. He remained without a visible patron, but he was nevertheless connected. He dropped the project to purchase the office of lieutenant general of Châtellerault, probably for the same reason that he broke with his father: he had associations among the discontented nobles. Cardinal Richelieu imprisoned some of them and executed the comte de Chalais as a warning to the rest. Descartes seems to have wanted to keep his head down, but his friends thought he was slipping off for secret meetings. When he did resurface, it was to join in assemblies where the latest ideas were being discussed, or to fight for the crown against England at La Rochelle, as did most of the nobility. He continued to remain on good terms with charismatic reformers within the church. But when he was finally flushed into the open, it was to raise questions about the principles on which one of the chief policies of Cardinal Richelieu was being erected, and he felt the ground slip from beneath his feet. Like so many who could not abide King Louis XIII's chief minister, Descartes would flee into exile.

Problems with the Cardinal

The conventional account of Descartes's period in Paris, between mid-1625 and the end of 1628, is that he lived quietly in order to begin developing his ideas about a "corpuscular-mechanical" philosophy associated with mathematics. The seventeenth-century biographer Adrien Baillet himself began the tradition of treating the later time in Paris as a period of quiet and reflection for Descartes, describing these years as more or less the makings of a philosopher-monk. Indeed, it is agreed by most commentators that by the mid-1620s, Descartes had solved some genuinely fundamental problems in analytical mathematics by using the equivalent of cubic and quadratic equations, and in 1626 or 1627 he developed his law of the refraction of light and began to apply it to the making of lenses. He was also drafting rules of reasoning.[1] The study of light was also replete with overtones about the quintessence from which the world was made, so some of his metaphysical views may have been shaped in Paris, too, building anew from materials left from the ruins of Aristotle and Galen and ongoing intense public scrutiny of Epicurean and hermetic philosophies. Baillet had already noted that Descartes had been pursuing his thoughts about *physique* from 1623, his Italian period. All these studies would provide the foundation for his later publications, which appeared from 1637 onward.

Descartes did not carry on his studies alone, however. He was able to pursue the work in optics with his previous acquaintance, Claude Mydorge, and others. Over time Mydorge would be best known for his work on "catoptrics" (the study of reflective surfaces such as mirrors) while Descartes would publish on "dioptrics" (the study of refractive bodies such as lenses); the first also published on parabolic curves while the second did so on hyperbolic curves. Apparently the wealthy Mydorge funded their work, and they employed Guillaume Ferrier as their talented artisan, who—following Descartes's own example as a fine master of the art of cutting glass—was able to produce better burning mirrors and magnifying lenses based on their studies. According to

Baillet, Jean-Baptiste Morin, an astrologer and projector who remained opposed to Copernicanism, helped Descartes secure instruments for his investigations, and so he may have been involved in the optical studies, too. Descartes's friend Étienne de Villebressieu—an alchemist, physician, and royal engineer—is another who may have joined in their experimental work at the time. The outspoken friar Marin Mersenne was then arguing that such studies in natural sciences would not only support a true understanding of God's ways but would also make prognostications more accurate.[2] According to Baillet, however, Descartes found that he was being led into abstruse sciences again and turned back again to "the study of man."[3] Baillet had said that, too, about Descartes's time in central Europe, and about his Roman period: it seems to be a broad hint about periods when human relations were consuming Descartes's attention.

Baillet drops further hints as well. The chevalier had not yet decided on a stable profession and therefore he insensibly fell into being nothing in particular, we are told—meaning, one assumes, that he found few opportunities for advancement at court or in government. Did he now have the wrong friends? After much deliberation, then, he decided to continue on his course, cultivating his reason and advancing truth to the extent possible. He had such contentment from this decision that nothing was sweeter to him than to close his ears to the world, Baillet says: this description in turns suggests that Descartes had decided to follow his conscience rather than make obeisance to authority. Baillet insists, too, that Descartes was also fortunate not to be a slave to the vicious passions that attract the young. For Baillet, Descartes kept to the piety he had learned at La Flèche and put more emphasis on living morally than on outward devotion—which we could turn around to mean that he cultivated an inward-looking faith, possibly even opposed to the new orthodoxy of ceremonial conformity being introduced to France by Cardinal Richelieu. And finally, Baillet says, Descartes was far removed from being a libertine.[4]

That last comment of Baillet's sounds especially defensive. It is an effort to distance Descartes completely from any earlier entanglements with libertinism in a period when such associations brought real danger. In December 1627, Richelieu would explain to the Assembly of Notables that libertinism had to be suppressed even more vigorously than Protestantism.[5] The execution of Giulio Cesare Vanini had begun a period of reaction. In 1624 three speakers had been exiled from Paris for planning to hold a meeting at which anti-Aristotelian propositions were to be proclaimed.[6] Théophile de Viau's trial occurred in September 1625, only a few weeks after Descartes's return. Baillet's comment might also be connected to his further insistence that a few weeks after returning from Italy, Descartes paid a visit to the royal court at Fontainebleau and again met Cardinal Francesco Barberini, to whom he recommended some of his friends, including Théophile's former friend, Guez de Balzac: Balzac had come under attack from Father Goulu, general of the austere order of the Feuillants and one of the most vocal of the reactionaries.[7] Descartes would continue to defend his friend energetically. But clearly the mood in Paris was rapidly shifting toward necessary conformity to Tridentine Catholicism, which was in turn shaped by neo-Aristotelian theology, and apparently Descartes and his friends were anxious. The era of Richelieu would not be easy for them.

What might Baillet have meant, then, about Descartes's return to "the study of man" in preference to more abstract lines of reasoning? We might assume that the life of his mind was occupied with many problems. But while Baillet continued to avoid direct discussion of the political world in which Descartes found himself, he dropped many pointed hints about his subject's continued involvement with the affairs of France.

After all, the political situation in France remained fraught. Richelieu continued to grow rapidly in power as he sought to bring to heel both the great nobles and the Huguenots. At the insistence of Marie de Medici, her son had in April 1624 procured her client a cardinal's cap and a seat at his council table, and Richelieu was offering advice and acquiring offices that would soon make the king entirely dependent on him. He had advocated the war in Italy because it served to undermine the ancient enemy Spain, but it offended Catholic *devôts* in France while also alienating Venice and turning the new Francophile pope into an adversary. At the same time, it gave the Huguenots an opportunity to make trouble while the crown was otherwise occupied, with the baron de Soubise seizing the isles of Ré and Oléron off the Atlantic coast near La Rochelle. The marriage in 1625 of the king's sister, Henrietta Maria, to the king of England, was a bright spot, tainted only by the Duke of Buckingham's embarrassing courtship of the queen herself, but while it wrapped up the dynastic strategy of Marie de Medici that went back many years, it, too, would soon bring further trouble for the cardinal.

In other words, both the successes and missteps of Richelieu were turning him into the chief focus of aristocratic discontent with the direction of the king's government. The Spanish-born Queen Anne was becoming dependent on Marie de Rohan, duchesse de Chevreuse, who was more and more openly Richelieu's enemy, while others were coalescing around Marie de Medici and her younger son Gaston. The old *politique* policies that had been meant to unite the country through law and negotiation were being discarded in favor of loyalty to the persons of the king and minister, while many of the aristocrats Descartes had once sought out were being sidelined or gradually driven into opposition.

Baillet notes Descartes's caution: "Seeing nothing that continued in the same state and condition" since he last resided in Paris, he did not commit himself to any faction, going along with whatever seemed right. Baillet also makes a comment that suggests the young cavalier was going out of his way not to give personal offense: "he lived in all appearance after the same manner as they do, who being out of employment, think of nothing else but to live a sweet peaceable innocent Life in the Eyes of Men." Descartes therefore made sure that his deportment had "nothing that smell'd of singularity in it": that is, he made certain that he did not become identified with those in opposition. In the end, Descartes was able to go about his "designs" without "producing ill effects in the Eyes or Imagination of others," so that no one "contrived to put an obstacle" in his way. He was successfully keeping his head beneath the parapet. But in a period of growing factionalism, neutrality also kept him publicly sidelined.[8]

But perhaps Baillet's comment about how Descartes turned away from abstract contemplations to "the study of man" meant that he was inevitably becoming involved with the factional positioning within France after all? Most of what Baillet conveyed about Descartes during these years came from the recollections of the friend of the family with whom Descartes made his visit to Poitou and Rennes in the summer of 1626, Nicolas Le Vasseur, sieur d'Étioles. In Paris, Descartes lodged with him, perhaps as early as his return from Italy. As receiver general of finances, Le Vasseur would have been working under a new council of two senior ministers for finance (*Surintendants des finances*), one of whom was Michel de Marillac, who later died in prison for supporting Marie de Medici against Richelieu. Since Descartes was himself unsuccessful in finding an office, Le Vasseur apparently engaged Descartes in his own work: Baillet comments that in arranging things with Le Vasseur, Descartes had "procured himself a kind of Settlement in Paris."[9] He was, after all, good with numbers. It might also have brought him into contact with Marillac himself or others of similar views.

Le Vasseur later passed on a report of some odd behavior on the part of Descartes. After returning from that visit to Châtelleraut and his father in the spring and summer of 1626, during the period leading up to the execution of the comte de Chalais, his guest moved out and took lodgings in the Fauxbourg Saint-Germain-des-Prés (named "little Geneva" in the later sixteenth century because of the high concentration of Huguenots there).[10] But some of his friends, such as Mydorge and Mersenne, told everyone where he was, causing many visitors to call on him, which forced him to return to Le Vasseur's. Were they naively trying to keep him thinking about natural philosophy, or keeping their friend from suspicion by making it impossible for him to remain incognito? Even more curious, after his return to Le Vasseur's, Descartes would disappear from time to time without notice or excuse and reappear days later just as suddenly, without explanation although with very polite apologies: Le Vasseur's wife, Descartes's hostess, was quite understandably irritated by this behavior, we are informed. Le Vasseur understood that Descartes had taken a room elsewhere in the city where he could meet with a select few friends in secret, but he knew nothing more.

Finally, after Descartes had disappeared for a considerable period, Le Vasseur was able to lay hands on Descartes's valet when they crossed paths in a market street, and he forced him to lead him to his master. Le Vasseur spied for some time through the keyhole, and although it was late in the morning, he saw Descartes lying abed and occasionally turning to a bedside table to write things down. This is the origin of the often-told story that Descartes did his best work while lying about. But one can imagine other reasons than philosophy for such unusual privacy and note taking, such as having been out all night, and not only for love. After watching for a while, Le Vasseur finally burst in and brought Descartes back to his house.[11] Was he, too, trying to keep his young protégé out of the clutches of Richelieu, whose nose for identifying possible disloyalty was acute? That might have implicated Le Vasseur himself.

The attentions of the cardinal were clearly worrying many of those who attended one or more of the several informal assemblies meeting in the city, sometimes called academies and other times salons. Madeleine de Scudéry, for instance, had to worry about the possibility of spies circulating through her salon, so she insisted that speaking of politics and religion was "taboo." Consequently "both salon habitués and academicians resorted to code names, metaphors, veiled allusions, well-placed silence, and other devices" to communicate their views.[12] The salon of the Protestant Marie des Loges, much frequented not only by Balzac but also by Gaston d'Orléans, became such a concern for Richelieu that he forced her to retire to the countryside in early 1629.[13] Catherine de Vivonne, marquise de Rambouillet, host of one of the most vigorous meetings of the period, received guests at her *hôtel* (urban palace) from her *lit de repos* (daybed—not unlike a modern settee or couch) in her *chambre bleue* (blue room): her salon is often described as a predecessor to the Academie française, the distinguished literary institution founded by Richelieu in 1635.[14] Yet Richelieu apparently worried about how her salon might be breeding subversion and sought to recruit her as an informant. "It was only through her connection to his niece, Mme de Combalet, that she was able to resist his orders." Coincidentally too, perhaps, the Madame de Combalet who intervened with Richelieu for Rambouillet—better known by the title she took in 1638, the duchesse d'Aiguillon—may have been the person that Descartes had in mind when he later thanked a lady for her help in getting his *Discours* approved for publication.[15] Rambouillet's *les précieux* were known in part for cultivating *honnêteté*, modeling the correct behavior that Descartes was imitating. Perhaps Descartes attended one of the salons with his friend Guez de Balzac, which would have given him the chance to cross paths with Gaston, a passionate reader of Honoré d'Urfé's huge and complex *L'Astrée* (1607–27), which contains many passages that interweave considerations of spiritual and carnal love in imitation of the ancient pagans.[16] "Monsieur" (Gaston) was certainly becoming a focus of discontent and a constant problem for the cardinal.

In Baillet's description of a regular weekly meeting at Le Vasseur's—which he claims was a proto-Académie française—he hints at the political sensitivities, writing that the group resisted being taken under the cardinal's protection.[17] The hint is strengthened when Baillet writes that the academy included "M. De Argues" of Lyon: this is Girard Desargues, a mathematician and engineer associated with the foundation of projective geometry who also played an important role in the siege of La Rochelle. Desargues apparently made Descartes known to Richelieu but, we are informed, although he acknowledged that he was much obliged, Descartes did not wish to gain anything from the introduction—a roundabout way of saying that Descartes had been noticed but did his best to keep his distance without giving offense.[18] Two other members of the *assemblée* (assembly) were associated with the cardinal, too: Jean de

Silhon, a philosopher who would defend Richelieu's idea of *raison d'état* ("reason of state"), and Morin, who would later attack Descartes's work. Baillet insists, however, that at the time Descartes and Morin remained respectful of each other.

But many more people in the group were not friends of Richelieu. One of the members of the group was a prince of the blood, the august Gaston Henri de Bourbon, son of Henri IV by his mistress Catherine Henriette de Balzac d'Entragues. He was bishop of Metz—a bishopric controlled by the house of Lorraine—and later became duc de Verneuil; Verneuil is known to have been deeply involved with the Chalais plot that the duchesse de Chevreuse helped organize on behalf of Gaston. In 1622 his sister married the son of the duc d'Épernon (Balzac's former patron), who would become one of Richelieu's enemies, too. Several members of the group were followers of the duc d'Orléans. For instance, there was a soldier-author whom Descartes had met at the siege of Gavi, Pierre de Boissat. Boissat was one of the gentlemen of Gaston's formal retinue. He was also a frequenter of Rambouillet's salon, a friend of Théophile in his last months, and later close to Théophile's lover, Jacques Vallée, sieur des Barreaux (who was also one of Gaston's gentlemen.)[19] "M. de Picot, prior of Rouvre," is better known as Abbé Claude Picot, later called Descartes's "agent concerning his domestick affairs"; he, too, was associated with Barreaux.[20] There was M. Jumeau, prior of Saint Croix and former tutor to Gaston. There was the intendant of the duc de la Rochefoucault, Jacques de Serisay, who also attended Gournay's salon;[21] his master, the duke, was later deeply involved with the duchesse de Chevreuse and Queen Anne. Of course, Descartes's good friend Guez de Balzac attended, and as we shall see, he would come under personal attack by the cardinal in later 1628.

A final set of members of Le Vasseur's academy are more difficult to place. Among them were Descartes's mathematical friends, Mydorge and Hardy, who were both members of the *parlement* of Paris, often at loggerheads with Richelieu. Another mathematician, M. de Beaune, sieur de Gouliou, was a *conseiller* (judge) of the *présidial* (tribunal) of

Blois, and yet an additional mathematician was the royal secretary, Jean Beaugrand, both of whom were undoubtedly loyal to the government, but for the moment the government included followers of Marie de Medici and Gaston as well as Louis. A little-known figure who would be close to Descartes for many years to come was Villebressieu, originally from Grenoble. Taken together, Baillet lists a relatively diverse group but one that on balance was aristocratic, libertine, and naturalistic in orientation, perhaps even a center of heterodox opinion—and distinctly leaning toward the party of discontent.

In addition, there were a few members of the group in religious orders who also probably had different agendas than the cardinal's. Guillaume Gibieuf, a doctor of the Sorbonne and priest of the Oratorian order, was said to be one of Descartes's "principle friends" in the academy. The Oratorians would later become powerful supporters of Descartes's views, and Gaston considered himself the protector of the order and chose his confessors from among them rather than the Jesuits.[22] (At the time, his confessor was Father Charles de Condren, who was deeply interested in alchemy.[23]) Outside of France, the Oratorians were a kind of voluntary association of laypeople and clerics who met together to share a charismatic faith through prayer, song, and teaching, having first been organized by Philip Neri, a Roman who modeled his life after the example of Jesus Christ and the apostles. Their aim was to revive "the primitive spirit of the early Christian Church," although like many other Catholic groups of Descartes's era, they also fostered the worship of the Virgin Mary.[24] In France, however, with the help of Marie de Medici, Father Pierre de Bérulle founded an Oratorian order that lived according to the direction of a superior-general (himself and his successors).[25] The Oratorians were in the forefront of reintroducing Catholic services to England, too, with Bérulle heading up a delegation of a dozen of them who accompanied Queen Henrietta Maria to London in 1625 to found her chapel.

If we look ahead to when Descartes published his philosophical writings, it is noticeable that many Oratorian principles were consistent with Descartes's metaphysics. French Oratorians emphasized not only the primitive church but the faithful intuition of God over doctrine, being especially negative toward Aristotelian-based rationalism with all its abstract terminology. As a part of their encouragement of lay piety, they urged ordinary people not only to follow their contemplative religious inclinations in their daily lives but also to engage in acts of charity. Bérulle was, for instance, instrumental in introducing the Order of the Blessed Virgin Mary of Mount Carmel into France in the charismatic form recently expressed by Teresa of Avila, while he also became the spiritual director of Vincent de Paul, who in turn organized the Dames de la Charité to help the poor and to ransom Christian galley slaves from North African states.[26] The duchesse d'Aiguillon had begun the process of becoming a Carmelite before being pulled out of the convent by her uncle, Richelieu; after his death, she devoted herself and her huge inheritance to De Paul and his ladies of charity. In terms of religious politics, the Oratorians composed a key part of a loose grouping called *les bons Français* (the good French), Catholic but Gallican reformers. The Oratorians would do more than any other group after the death of Descartes to spread his reputation as an important philosopher, excepting only some among the Jansenists, who in turn were early on associated with the Oratorians.[27]

Mersenne was not mentioned. Perhaps he only became well acquainted with Descartes shortly before Descartes departed from Paris. Mersenne had taken vows as a Minim friar, which meant an ascetic life that included abstinence from meat and dairy products and going barefoot or in sandals (discalced). The order included such people as the charitable mystic Francis de Sales (later sainted by the church), who promoted the adoration of the Virgin Mary and consulted with Bérulle about the Carmelites. But Mersenne was also associated with the duc d'Orléans, having dedicated his 1625 book (*La Vérité des Sciences*, or *The Truth of the Sciences*) to the prince. Its agenda was to use the new sciences, particularly those rooted in physical and mathematical studies, to support a verifiable natural philosophy. Its dedication identifies Gaston as the tutelary *génie* behind the savants his day. Gaston's "true science" (*science véritable*: Mersenne's phrase) included not only alchemy but the use of mathematical compasses for the study of astronomy, fashioned by the excellent instrument maker Ferrier, with whom Descartes also worked.[28] Perhaps it is only a coincidence that Le Vasseur reported Descartes as dressing in green taffeta, the color in which Anne of Austria and Gaston d'Orléans dressed their servants?[29]

Perhaps it was also through these circles that the court artist, Simon Vouet, created a portrait of Descartes (as recently discovered by Alexander Marr; fig. 11).[30] Vouet did much work for the Oratorians.[31] He had recently acquired a new method that allowed him to sketch with pastels.[32] Vouet may have met Descartes in Rome, where he was on good terms with the Barberinis and no doubt on similar terms with the connoisseur Béthune as well, but he had returned to Paris in late November 1627, giving him a chance to capture this image of Descartes over the winter—he depicted many young courtiers around the king. Another member of Le Vasseur's academy was "M. Sarazin," who is probably the sculptor Jacques Sarazin, who returned to Paris from Italy in 1628 and married Vouet's niece. (He would later execute a bust of Gaston d'Orléans as Hercules.[33]) In Vouet's portrait, Descartes is captured with hair that flows down to the length of his jaw, a bit shorter than in his later portraits. Since the Italians wore much shorter hair, perhaps Descartes had just begun to grow it out to the shoulder length of French courtiers (and Dutch patricians) that one sees in his later depictions. He also has a moustache with upturned ends and a beard on the lower lip and chin clipped short, also common among courtiers and their imitators, which he would keep to the end of his life. Vouet's depiction of his mouth makes him look pleased with himself and the world, but the eyes are full of melancholy.

Other hints about Descartes's personal circles in Paris suggest that he continued to keep up relationships with even less conventional figures, too. A letter of Balzac to Descartes from March 30, 1628, asks him to please remember that his friends in Paris are awaiting "*De l'Histoire de vostre Esprit*," something that might now be called an autobiographical reflection.[34] Descartes had promised Balzac this account in the presence of Père Clitophon, called in ordinary speech M. de Gersan. Gersan has

Figure 11. René Descartes, c. 1628, as identified by Alexander Marr, in *Head of a Man Wearing a Collar.* Pastel sketch by Simon Vouet, Musée du Louvre. © RMN-Grand Palais / Art Resource, New York.

been identified as François du Soucy, sieur de Gersan or Gerzan, a novelist who wrote two alchemical treatises and whose interest in hermetic philosophy has led some observers to suspect him of Rosicrucianism.[35] Moreover, according to a letter of 1631, Descartes's friend Villebressieu had been conducting experiments on metals for twelve years, a line of inquiry associated with alchemical interests; in a written comment commending Villebressieu's work, Descartes suggests that he himself had participated in the trials as well.[36] He had apparently not given up his curiosity about the occult philosophy.

The Campaign for La Rochelle

The summer of 1626 may have been disappointing for Descartes given the sad business of Chalais, the apparent rupture with his father, and his setting aside any interest in the lieutenancy. It was apparently followed by the strange absences from Le Vasseur's lodgings that suggest clandestine meetings. The late summer of 1627 would be even busier, since Descartes would be involved in some of the opening actions related to the siege of La Rochelle.

Borel, Baillet, and other early biographers insist that Descartes took part in the campaign, and although more modern authors have expressed doubts, it seems to be true.[37] The discovery in the early twentieth century of the journal of his Dutch friend Isaac Beeckman places Descartes in the city of Dordrecht on October 8, 1628, before completion of the siege, so doubts are understandable.[38] But they seem to be dispelled by an entry probably referring to Descartes in Richelieu's papers, previously unnoticed, dated November 8, 1627: "Le cap^{ne} Descart" is listed among the captains of a fleet of ships sent to relieve the royal citadel of Saint Martin de Ré.[39] In fact, this "Descart" is listed as second in command of the exploit, the *visadmiral* (vice admiral). If this is our man, then he certainly took part in the campaign.

The entry refers to one of the early events related to the struggle for control of the great Atlantic port, which involved the battle for the offshore Isle de Ré that controlled the approaches to La Rochelle's harbor. An explanation for those events best begins with the Duke of Buckingham, who a few years earlier had accompanied his friend Charles, Prince of Wales, on a journey in pursuit of royal marriage with the ruling family of Spain. On the way to Madrid, they had stopped in Paris and been impressed by the ladies of the French court. After plans for a Spanish marriage were abandoned, a union was negotiated between Charles and the sister of the French king, Henrietta Maria. With Charles's plans interrupted by his father's death and his own ascendancy to the English throne in 1625, King Charles I married his bride by proxy outside Notre Dame Cathedral—his cousin Claude de Lorraine, duc de Chevreuse, standing for him—and sent Buckingham to escort her and her large retinue back to London. While in France, Buckingham also engaged in some diplomacy, proposing an open alliance against Spain, but Richelieu was not prepared to undertake more than he could, and he declined. With the duchesse de Chevreuse's complicity, Buckingham also engaged in his affair of the heart with the French queen, which led to their mutual embarrassment.

Earlier in the year, the French crown had faced further militancy on the part of French Huguenots led by Benjamin Rohan, baron de Soubise, who seized the isles of Ré and Oléron off La Rochelle, itself a Protestant citadel. With Dutch and English naval support, the French were able to retake the islands[40]—that was part of the campaign of the summer and autumn of 1625 in which Descartes seems to have been at least an onlooker—and Richelieu started to rebuild the French navy with purpose. For his part, encouraged by his new lover, the duchesse de Chevreuse herself (who had accompanied Henrietta Maria to London and spent some weeks there afterward), Buckingham's wounded pride about the encounter with the queen quickly turned against the cardinal. As the English Lord High Admiral, Buckingham also thought more grandly about how to strike at France's fledgling navy, and Soubise had fled to England, giving him high-placed leverage. Secret plans were forwarded for a new alliance against the French regime, with Savoy, Lorraine, and Venice cooperating with Spain, all headed by Charles I; but Spain signed the Treaty of Monzon with France, and Richelieu held onto his position despite becoming the focus of opposition, with Chalais's lost head one of the warnings of what the cardinal would do to keep his place.

In the summer of 1627, then, Buckingham embarked on an expedition to liberate La Rochelle from France, expecting that the Huguenots would rise in his favor and that Savoy, Lorraine, and other powers would then bring their own forces into the conflict after an initial victory. He landed forces on the Isle de Ré, pushing the French into the fortress of Saint Martin at one end of the island, to which the duke laid siege. But his move would have further consequences. Across the bay, in La Rochelle itself, Soubise was allowed to enter and speak. He stirred his audience, and with potential English allies on hand, the Rochelais began to make demands on Louis XIII. In September they went so far as to fire on royal forces under the command of the duc de Angoulême—one of Descartes's possible former patrons—who in turn responded by initiating a siege of the city. It would not end well for them. Another possible object of patronage for Descartes, Gaston d'Orléans, quickly arrived to take command of the relief of Ré, only returning to Paris in mid-November when the king and cardinal had arrived and the action was concluded in Gaston's favor.[41] The dashing Buckingham had provoked a chain of events that would end his own life and bring the Huguenots of France to their knees.

On the Isle de Ré itself, the French royalists defeated the duke.[42] On the night of October 7–8, a small fleet of supply ships commanded by Claude de Razilly managed to slip through the English blockade to bring supplies to the garrison of Saint Martin, allowing it to avoid capitulation. Two weeks later, they also managed to land a large number of troops elsewhere on the island under Henri de Schomberg. Buckingham responded on October 27 by trying to take Saint Martin by direct assault but failed (the engineers had miscalculated the height of the walls and the scaling ladders proved to be too short). A few days afterward, the duke began his retreat. Attacked in turn by the French, he lost many of his remaining men and was forced to withdraw in disgrace. He was assassinated a year later. "Le cap^{ne} Descart" therefore either took part in the effort to bring critical supplies to the starving and exhausted forces in Saint Martin, or he helped in a venture to resupply the city or to see off the retreating English, having been afterward mentioned in dispatches.[43]

Is the entry in Richelieu's papers confirmation that the Descartes

Figure 12. Siege of La Rochelle, detail; Harbor Barrier below. Engraving by Jacques Callot, Musée du Louvre, Paris. Erich Lessing / Art Resource, New York.

who took part in the fighting was René? Perhaps it might refer instead to his father or one of his brothers. If his father had been involved in the affair of the Spanish spy, then he certainly exhibited an ability to move quickly and decisively, and eight months after the relief of Saint Martin, he received a letter of honor from the king, possibly implying a link. But the siege took place twenty years after the espionage, by which time Joachim was well over sixty years old. More important, he is not known to have had any military experience. Nor, as far as we know, did any of René's brothers: Pierre and Joachim II were or would be members of the *parlement* in Rennes, and Joachim II was only in his early twenties, with two other brothers by their father's second marriage even younger. Even if they were inclined to contribute to the fighting, then, it is hard to see how they might have been made second in command of a naval unit. But René had considerable military experience, was in his prime at just over thirty years old, and was still seeking a place in the world. Likely noble patrons of previous years were commanding the forces. At the time, officers with fighting experience might as easily be employed on water as on land. Baillet himself insists that Descartes took part in the relief of Saint Martin.[44] It is likely, then, that the name of the vice admiral belongs to him.

He may well have remained involved during the rest of the year-long siege, at least from time to time. Baillet reports that he returned in August 1628, and he could not bring himself to leave the camp until the conclusion of the business. He "again" obtained "the pleasure" of conversing with the engineers, in particular his friend Desargues, who played an important part in the planning and execution of the siege.[45] Given his background as a military engineer, Descartes may well have been particularly interested in the construction of the siege works at La Rochelle. Most everyone was. Richelieu had decided on a plan that would cut the city off from the sea by erecting a huge barrier in the harbor just beyond cannon-shot. While the plan almost bankrupted the kingdom, the engineers succeeded and caused the failure of all attempts at succor from the sea. When an expedition by an English relief fleet in September 1628 managed to engage the French but failed to break through the seawall, Descartes was there. A short truce was agreed and negotiations begun. During the ceasefire, gentlemen from the opposing armies toured each other's forces, and Baillet insists that Descartes was among the Frenchmen who visited the English fleet. As for Descartes's recorded visit to Beeckman in Dordrecht early in October: the Dutch government had supported its French allies during the siege by sending them troops and supplies by ship, but there was vocal and sometimes violent opposition from many Calvinists who sympathized with the Rochelais. By then Gaston d'Orléans had rejoined Louis XIII in the camp. Was Descartes sent to assess Dutch support for the continuation of the siege at a critical juncture, or simply helping to transport the proffered aid? When Descartes left Beeckman, he promised to send him a manuscript from Paris, suggesting that was where he was headed.[46] He was still on the move.

The English had dragged the city into a fight it could not win, and with more than 80 percent of the original population dead from famine, disease, and violence, the mayor was finally forced to surrender unconditionally. The French king graciously did not sack his city. According to Baillet, Descartes entered with the royal forces of occupation on October 28, 1628, and participated in the solemn processions of the holy sacrament through the streets on November 3 to memorialize the dead. The king was detained in La Rochelle until November 11 by an attack of gout; according to Baillet, Descartes returned to Paris at about the same time, arriving by Saint Martin's Eve (November 10–11, a date often associated with significant events in Descartes's life.)[47] The victory consolidated the power of the monarchy and of Catholicism in France.

Confrontation and Departure

But then something even stranger happened. By March 1629, Descartes had moved for a second time to The Netherlands, which he would treat as his home for the next twenty years. He had returned to Paris in November from La Rochelle to his many friends, an accomplished and loyal thirty-two-year-old soldier and diplomat as well as an excellent mathematician and experimentalist, by now known to some of the great French aristocrats and to royal officials as well as papal ones, and an associate of many of the Parisian *esprits forts* (strong spirits). He was nearly at the peak of his powers. The subsequent departure for the Dutch Republic therefore comes as a surprise. Descartes seemed to have everything going his way only to take to the road yet again. It must have been either a deliberate and voluntary move or something that could not be helped, such as involuntary exile. What happened?

Baillet and almost all subsequent authors chose to downplay Descartes's move, explaining that it was by choice, intended simply for the purpose of finding a quiet place to write. The usual story is as follows: Descartes had returned from Italy in 1625 and became associated with an intellectual circle around Father Mersenne. Then, in 1628, Descartes attended a lecture by the alchemist M. de Chandoux at the nuncio's palace, at which the nuncio, Mersenne, Cardinal Bérulle, and many others were assembled. At the conclusion of that meeting, he himself was asked to speak, and in doing so he showed that Chandoux's way of thinking was incorrect whereas his own new philosophy was based on principles demonstrably true, following which everyone encouraged him to publish. After the meeting (the current story goes), "according to Descartes," Bérulle granted him a private audience and "encouraged him to develop his philosophy as an antidote to atheism." But "Paris was not allowing Descartes uninterrupted time to work on his various projects and even the French countryside did not provide him enough peace and quiet." Having been invited to write, and seeking a place to do so, he left for The Netherlands.[48] But there remain many questions about such a simple scenario.

The Meetings

The consensus account makes Descartes's public confrontation with M. Chandoux and a subsequent private conversation with Cardinal Bérulle the critical turning points, following which he determined on a life course that required peace and quiet to write down his ideas. The account of both events and the subsequent decision originates from Descartes's first biographer, Borel, as amended and corrected by Baillet, whose further details again hint at more than he says explicitly.[49] We can break the original story into three parts, which continue to be represented in all later explanations: an *assemblée* of important people at which Chandoux and then Descartes spoke, an interview with Bérulle, and a decision to depart.

Before a closer examination of the further accounts, however, let us note that the best remaining evidence of the assembly comes from undated fragments of a letter Descartes wrote to his close friend Étienne de Villebressieu, from Amsterdam, dated by Adam and Tannery to 1631 and often assumed by others to be from the summer of that year.[50] Descartes and Villebressieu had gone on a journey to Denmark, but Descartes had returned in advance to "our lodgings" at "The Old Prince" in Amsterdam, where he was healthy and looking forward to Villebressieu's return, and from where he wrote. In the opening paragraph, however, for some reason Descartes feels the need to remind Villebressieu of "these two results of my fine rule or natural method" that emerged from the discussion "that was forced on me" in the presence of the papal nuncio, Bérulle, Mersenne, "and all that great and learned company assembled" to hear M. de Chandoux "lecture about his new philosophy." Descartes reminded Villebressieu that he "made the whole company recognize what power the art of right reasoning" (*l'art de bien raisonner*) has over minds "who have no learning beyond the ordinary," showing how much better his own principles were.[51] Everyone had been convinced that he was right, and Villebressieu and the rest had begged him to publish.[52] Let us set aside, for the moment, the interesting ideas that are being discussed between the two in the document and simply note that this fragment is probably the chief source for the later accounts about the meeting in Paris.

The written fragment contains no date, nor does Descartes mention one for the meeting. In Baillet's version, however, the assembly took place at the end of 1628, a few days after Descartes's return to Paris from La Rochelle, which is likely enough.[53] Since the *assemblée* was hosted by the papal nuncio, Giovanni Francesco Guidi di Bagno, known in France as de Baigné, it cannot have been earlier than 1627. Bagno was one of the closest confidantes of Urban VIII, returning to Paris as his representative in May 1627 and taking up residence in the Hôtel de Cluny.[54] The landing of Buckingham on the Isle de Ré that began the La Rochelle campaign occurred in September, after which it is likely that Descartes was involved in the campaign, certainly during the first autumn and probably from time to time thereafter. The *assemblée* also included many eminent figures who would usually have been accompanying the royal court. The meeting could have occurred before the siege, in the summer of 1627, but if it were afterward, it would have been more likely to have been after the campaign, at the end of 1628—unless it were March 1628, when the king and his retinue returned briefly to Paris. Later that year, following his entrance into La Rochelle and period of recovery from his gout in early November, the king slowly made his way back to Paris via royal hunting lodges and palaces, finally entering to great celebration on December 23. Descartes had returned by November 11, Baillet says. Like him, many courtiers and officials would have gathered in Paris between then and Christmas and, with the king still away, not yet have picked up their full responsibilities. It may well have been then that the nuncio saw an opportunity to act as host for an invited discussion.

Even though we are given the names of only half a dozen people, we should probably imagine a substantial gathering—what Descartes later called a "great and learned company assembled"—rather than a small group. Much of the most recent evidence of Chandoux's own views has been extracted from a manuscript copy of some of his writing that is stamped with the coat of arms of Philippe de Béthune, whom Descartes knew: although Béthune himself remained in Rome until May 1630, perhaps Béthune's people were among the participants.[55] The discoverer of that document also thinks that the person who might have persuaded Bagno to host Chandoux's presentation was the Oratorian priest Charles de Condren, confessor to Gaston d'Orléans.[56] In the 1630s two alchemical books would be dedicated to Gaston (known familiarly as "Monsieur"), so perhaps Chandoux meant to address Gaston's circle, a group enlarged by other possible discontented nobles?[57] If so, perhaps Monsieur himself attended.

Descartes was probably invited because he and Bagno were previously acquainted. Bagno had first come to France in 1625 as an assistant to Cardinal Barberini, in whose retinue Descartes began his own return; Descartes might even have met Bagno earlier, since the papal nuncio was the older brother of the Cardinal Bagno whom Descartes had met in the Valtellina. The younger Bagno had a soft spot for courteous French intellectuals, even people known as libertines, since in 1630 he appointed Gabriel Naudé as his librarian and later used his influence to obtain a post for Naudé as a French royal physician. (Later in life, Naudé also became librarian to Cardinal Francesco Barberini, Cardinal Mazarin, and Queen Christina.)[58] Naudé had studied medicine with Cesare Cremonini in Padua and developed into a notorious freethinker, explicitly refusing to believe in miracles, for instance. Bagno and Barberini were among the princes of the church who imagined that a restored universal faith would be aligned with a full and correct understanding of God's creation. Descartes was well enough known as a reformer to be considered by another nuncio in 1632–33 for a chair in theoretical medicine at the papal university in Bologna.[59] Baillet was of the opinion that if Bagno had been in Rome at the time, the cardinals would not have voted to condemn Galileo.[60] Knowing Descartes from earlier days, Bagno himself may well have invited Descartes to the meeting he hosted.

But perhaps the other great person who was identified as present also wanted to see how Descartes would respond: Cardinal Bérulle, personal adviser to both King Louis XIII and Cardinal Richelieu but increasingly known as a loyalist to Marie de Medici. Descartes may well have met him previously. Bérulle had entered the royal court in 1599 as an almoner, then became preceptor to the Dauphin Louis (born in 1601), then rose to the place of Marie de Medici's spiritual director about the time of the assassination of Henri IV. The queen regent's support was undoubtedly critical in his founding of the Oratorians in 1611, and he in turn increasingly acted on her behalf in personal and international diplomacy. He had been a key participant in the 1619 conference in Angoulême that first reconciled mother and son (where he would certainly have met Balzac). Bérulle also managed to obtain the papal dispensation in 1624 that allowed the marriage of Henrietta Maria to Charles of England—dear to the queen mother's policies—and he accompanied the new queen of England to London. After his return to France in September 1625, he handled the negotiations in Paris over the Valtellina with the nuncio Barberini. Bérulle would go on to negotiate the Treaty of Monçon/Monzon with Spain (March 1626) that sorted out the conflict in the Valtellina; partly for that, he was granted a cardinal's cap in August 1627. He and Marillac (for whom Descartes's host, La Vasseur, worked) led the opposition to Richelieu on the king's council, and they originally opposed the La Rochelle campaign, preferring a Franco-Spanish alliance against England; but like everyone else, Bérulle was present during the siege.[61] There were many recent occasions on which Descartes was likely to have been in the presence of Bérulle. It is even possible that Descartes had quietly been working on behalf of Marie de Medici through Bérulle for many years.

The meeting had been called to hear from a person until recently little known to historians, "Chandoux."[62] All Baillet knew about him, through Claude Clerselier, is that he was executed in 1631 for counterfeiting.[63] Yet Baillet is also clear that Chandoux had wormed his way into the confidences of some of the most eminent men of the period. Thanks to the recent investigations of Sylvain Matton, we now know that Chandoux was from the same social stratum as Descartes himself: *ecuyer* (squire) Nicholas de Villiers, sieur de Chandoux.[64] He and a colleague, Robert le Toul, sieur de Vassy, *conseiller du roy* (official adviser to the king) from Avallon (in Burgundy), had lived under the same roof in Paris and, following the Rosicrucian scare, had been imprisoned on charges of helping to promote Rosicrucian views, and for being magicians, Pelagians, and atheists.[65] Chandoux, at least, spent two years in confinement before being released in late 1626 or early 1627, and he was clearly trying to reestablish himself in elevated social circles where ideas about nature were being discussed.[66] He was at the time proclaiming that he had found a method to make potable gold, a universal medicine that had long been sought by Paracelsians and alchemists. A meeting had been arranged to hear his new hopes for the reform of medicine and natural knowledge. But why?

Richelieu himself was just then beginning to establish institutions that would both encourage and regulate the growing number of people engaged in the "new philosophy," and he favored alchemically inclined physicians. Among the policies put forward by Richelieu to an Assembly of Notables in the winter of 1626–27 were educational reforms meant to strongly encourage more technical and commercial teaching and innovation.[67] He also seems to have considered the invention of new remedies to be especially useful for the kingdom. Richelieu had already granted royal permission (in 1626) for an alchemical physician, Guy de La Brosse, to found a garden in Paris where he would teach chemistry and botany, but the Faculty of Medicine—which was associated with the Sorbonne and had a long history of refusing to acknowledge that any good could come from chemical medicine—objected.[68] It would take more than a decade before the *Jardin du roi* (Garden of the King, now known as the *Jardin des plantes*) was opened.

In the same year, another of Richelieu's loyal servants arrived in Paris to help his patron: Théophaste Renaudot, a most ingenious person with a medical doctorate from Montpellier who also favored chemical medicine. Although originally a Huguenot, Renaudot had made a favorable impression on Richelieu early in his period as bishop of Luçon, when he helped the bishop work out plans for aiding the poor. As a reward for his efforts, in 1612 Renaudot obtained one of the posts of royal physician, and the office was confirmed through Richelieu's explicit efforts in 1617, following which (in 1618) Renaudot became the Comissaire Général des Pauvres du Royaume (the royal official overseeing the poor). He also served as personal physician to Abbé Michel le Masle, sieur de Roches, Richelieu's confidant and secretary. In 1626 he converted to Catholicism and moved to Paris, shortly thereafter publishing two pamphlets for his patron, one dedicated to Richelieu, the other praising him as "the Gallic Hercules." More recently, on March 31, 1628, Louis XIII had confirmed all the previous patents granted to Renaudot while also giving him permission to establish his Bureau d'Adresse. The bureau was certainly operating by 1630, but it may have been in existence a year or two earlier. It was established to solve urban poverty by turning beggars into self-sufficient workers, and by putting employers in touch with potential employees mainly by offering "a public registry of goods and services for buyer and seller," thereby freeing up the labor market, which was otherwise controlled by various guildlike companies. The bureau also acted to stimulate innovation by publishing one of the earliest newspapers, the *Gazette*, and by hosting and publishing (and anonymizing) weekly public meetings and debates, in French, on all kinds of subjects, including new findings in the arts and sciences; the only prohibited topics were politics, religion, and the reputation of persons.[69]

In other words, Richelieu was trying to tackle problems of poverty through new methods of invention and information exchange, and by encouraging the development of new projects by alchemical physicians, even establishing new institutions to carry out the work; Chandoux had ideas about how to make a famous panacea; he was invited to present his views to the great and the good. Chandoux had already obtained access to many of the great nobles—perhaps including Richelieu himself?—and Bagno had invited along a great number of the savants and *beaux esprits* as well as other senior clerics to hear from him. Perhaps the nuncio was seeking out an alternative projector to Renaudot, with Chandoux as the chief candidate? But someone with a similar background to Chandoux who had contrasting ideas about the constitution of nature had been invited, too, and would speak up: René Descartes.

Confrontation and Conversation

Descartes brought Mersenne and Villebressieu along with him to the *assemblée*, possibly because he was expecting a public encounter. But he does not seem to have sought a confrontation. In the letter to Villebressieu, Descartes wrote that the discussion had been "forced on me." He had, after all, previously avoided controversy, whatever his personal sympathies. But now he was required to speak his mind in the presence of powerful people, and his words suggest that he might later have regretted it.

According to the later reports, Chandoux spoke well and completely undercut the philosophy of the scholastics before proposing his own alternate theory. A recent analysis of his philosophical ideas proposes that he argued for a theory of mixtures based on ideas about how the elemental principles were rooted in qualities, which could be manipulated in ways that would transform those qualities and so the substances of which they were a part. For Chandoux, qualities were primary, matter secondary; for the Epicureans and Galileans, matter was primary, qualities secondary. Chandoux's position was in keeping with some of the latest hermetic theories about alterations in nature associated with transmutation.[70] He was received with almost universal applause.

But Descartes did not make the usual signs of satisfaction. Cardinal Bérulle noticed Descartes's silence and demanded that he comment. In reply, Descartes deferred to the judgment of those in the room who were more knowledgeable than he. But Bérulle seems to have taken that response to be dissimulation and pressed him for his views, supported by Bagno and others. Having been asked to speak his mind clearly, Descartes began by politely praising much in Chandoux's speech—Descartes, too, was interested in improving medicine and industry, and deeply interested in chemical operations.[71] But because he was critical of the qualitative approach of both scholastic and alchemical theory, he criticized Chandoux's method of establishing truths about nature. Descartes then demonstrated that, in a series of twelve steps, each of which could be shown to be correct, he could turn each of Chandoux's propositions on its head; in a similar way he could show that anything taken by him to be false was true. The audience was impressed not only by Descartes's arguments but even more by the force with which he made them: he had presence. They asked him for an alternate way of establishing the truth, and he explained that he had found an approach based on mathematics that could prove universal laws.

We can infer that Descartes's objections were based on the corpuscular materialism that he seems to have had in mind already. In the letter of 1631 to Villebressieu, Descartes expressed the opinion that his friend should publish his findings as a set of propositions "so as to force somebody else to supply them with research and observations" if they wished to object to any of them. That suggests something like the rules of reasoning almost all commentators agree he had sketched out in 1628, known as *Rules for the Direction of the Mind* (*Regulae ad directionem ingenii*). But the *Rules* appeared in print only posthumously, first in Dutch (1684) and then in Latin (1701). Richard Serjeantson has recently discovered an early manuscript version that was circulating privately, which may tell us more about Descartes's early thoughts about his rules, but for now an intensive study of them by John Schuster suggests that Descartes "abruptly abandoned" work on them around 1628. The early parts suited his corpuscular and mathematical approach to analyzing nature—it is materialistic—but the more ambitious later rules, Schuster thinks, ran into metaphysical difficulties that could be resolved only by philosophical dualism; at that point he abandoned the project.[72] Perhaps in the public confrontation at the meeting Descartes spoke about positions similar to those laid out in the early parts of the *Rules*.

As for the goal of improving life, like De la Brosse, Renaudot, Chandoux, and many others, Descartes placed the greatest weight on medicine. A few years later, in his first book, the *Discours*, he stated that his chief aim was "discovering a practical philosophy" that "would be very useful in life." That was reiterated in a letter to Mersenne in February 1637, where he described his book as consisting "much more in practice than in theory." His method would lead above all else to "the maintenance of health, which is undoubtedly the chief good and the foundation of all other goods in this life." Descartes declared that "I have resolved to devote the rest of my life to nothing other than trying to acquire some knowledge of nature from which we may derive rules in medicine which are more reliable than those we have up to now." The kind of issues about health and medicine that had been brought up in the confrontation with Chandoux clearly continued to motivate him in later years.[73]

Descartes had obviously prepared well for his public moment, leaving a strong impression on his audience. According to Baillet, Bérulle (fig. 13) was so moved that he asked to hear it all again.[74] Descartes would have understood that such a request from someone like Bérulle was not only an honor but an order, and he called on him a few days later. He then explained to the cardinal that the application of his method would greatly benefit medicine and mechanics, leading to the conservation of health and the relief of labor. Impressed, Bérulle told him that he had an obligation to explain his views on paper, which would succeed with God's continued help.[75]

Let us note that Baillet does not say that the cardinal invited Descartes to enter into any discussion of theology, much less to present his views either as an antidote to atheism or as helping to establish a truer foundation for religion.[76] Bérulle would certainly have reserved such kind of religious matters to professionals like himself, and in later years Descartes always deferred to authorities on matters of doctrine. If Bérulle had looked to Descartes to enter into religious discussion, he might have invited him to have further conversations with one of Descartes's friends, a member of the *assemblée* at Le Vasseur's, Father Gibieuf, who was an Oratorian and so under the cardinal's direction. Gibieuf was then working on his book *De libertate Dei et creaturae* (*On the Liberty of God and His Creation*), where, it has been said, Descartes found many of his religious principles.[77] But according to Baillet, Bérulle simply gave Descartes his blessings for his attempt to develop a new philosophy that would relieve some of the sufferings of human life, which in Bérulle's mind indicated that some form of divine guidance was helping Descartes.

Figure 13. Cenotaph of the heart of Cardinal Pierre de Bérulle. Sculpture by Jacques Sarazin (acquaintance of René Descartes), Louvre Museum, Paris. © RMN-Grand Palais / Art Resource, New York.

Becoming a Sectarian

We now come to the third element in Baillet's narrative, the subsequent decision to remove himself to another place. Baillet says that Bérulle's encouragement made an impression on Descartes, since it agreed with what his own nature and reason urged him to do. Until then, he had not had enough confidence in his views and had avoided being identified with any "sect."[78] But Descartes wanted to keep his liberty—implying personal danger if he made a misstep—and so he wished to work directly with nature rather than as a member of anyone else's school of thought or in response to any previous opinions.[79] In other words, he wished to go his own way rather than to engage in quarrels with those in power. Descartes's friends were also said to have redoubled their demands that he publish. So he considered the means to accomplish this end, seeing only two obstacles: the heat of the climate and the crowds of the world.[80] He therefore sought complete solitude (*solitude parfait*) in a country of average coolness (*dans un pays médiocrement froid*), where he would be unknown (*où il ne seroit pas connu*). He went north to the Dutch Republic, leaving Mersenne to manage his communications via letter (*commerce de lettres*) and Picot to manage his "domestic affairs."[81]

Let us begin to decode this third aspect of Baillet's narrative by noting that it gives weak reasons for Descartes to leave France. We have already noted the many ties that bound him. Bérulle was simply adding his encouragement to that of Descartes's friends, who wanted him to write down his views; none of them urged him to leave. But to write down what? He had apparently abandoned the project on rules of reasoning. Perhaps he was working on "*de l'Histoire de vostre Esprit*" that Balzac and his friends were anticipating in the spring of 1628? That was said to contain Descartes's attack on the scholastic "giants" of the day and also "your various adventures in the world and in the higher regions of the atmosphere" (*vos diverses aventures dans la moyenne et dans la plus haute region de l'air*).[82] One might construe that as an invitation to Descartes to put his "highest" ideas on paper, but it might also be an invitation to write up his memoirs of recent travels in Europe, or even to finish up a work of fiction similar to Cyrano de Bergerac's accounts of the histories of states on the moon and sun. Whatever it was, it has never come to light: Descartes thoroughly suppressed it (unless his "dream" episodes were part of an early sketch of an opening scene).

Any writing of the kind could ordinarily have been completed in town. Other authors wrote in the mornings and then engaged in visits to the salons in the afternoons. Moreover, the optical experiments Descartes had begun with Mydorge required dedicated space and the assistance of Guillaume Ferrier, further linking him to Parisian places. (Descartes later unsuccessfully tried to attract Ferrier to work with him in The Netherlands.) Of course, his writing would have been done in the midst of a social whirl, and there is evidence of periods when Descartes sought secrecy. But there is no evidence to connect his moments of secrecy with his philosophy (or anything else, for that matter—they remain mysterious and intriguing).

In other words, the meeting at the papal nuncio's apartments had simply brought Descartes's position into the public arena. All the incident explains is how he was further encouraged: or to put it in modern terms, the incident should have given him confidence in expressing himself among the leading figures in Paris. Neither the meeting of the *assemblée* nor the visit with Bérulle explains why he would leave the city.

Only two hints about possible reasons are offered by Baillet, and both are metaphorical. We are told that Descartes wished to remain free of identification with any sect. That might have been difficult enough in a city awash with intense conversations layered with innuendo and implication; but now he had been flushed out of the shadows, where every hint about his opinions would be open to scrutiny. What "sects" are we talking about? And why would joining one or another threaten his liberty? Baillet's comment about his need for "perfect solitude" is not a flat statement about a desire to live apart from other humans, then, but a suggestive metaphor that underlines Descartes's attempt to keep his independence from any particular party, remaining unentangled. Similarly, the comment about the climate is very odd. According to Baillet, we are in darkening late November, in Paris, a month away from the shortest day of the year, and we can imagine the bare limbs on the trees and the winds and rain out of the north deepening the mud in the streets and roads. Why was it not cool enough, and why does a move even farther north sound like a solution? Only the metaphorical heat of the moment can point to his meaning.[83]

To get to what might have been making the moment "hot," however, we need to consider the current controversies in the kingdom once more, which again revolved around events in Italy, where the climate was certainly warmer. The occasion was prompted by the death of the Duke of Mantua at the end of December 1627; the French duc de Nevers, who was also from the Gonzaga clan, declared himself the rightful successor and arrived almost immediately to take possession of the city and territory. Other claimants, however, were Ferrante II, Duke of Guastalla (supported by Emperor Ferdinand II) and Prince Charles Emmanuel of Savoy. Charles Emmanuel's son, Victor Amadeus, was married to Christine—whom Descartes had met previously—and her mother, Marie de Medici, strongly supported the case for Savoy and carried most of the king's council along with her, including her advisers Cardinal Bérulle and the marshal and count Louis de Marillac. But Richelieu considered that the interests of France lay in the duc de Nevers, and he convinced Louis XIII to support his own position. Richelieu's differences with Marie de Medici over Mantua at last convinced her that he, formerly one of her creatures whose career had been made by her preferment, had shifted his personal allegiances and was now using the king against her for his own ambition. The royal government was divided deeply, again.

The astute Marie de Medici might have been noticing Richelieu's changes of loyalty for some time. After all, the discontented had begun to plot against him in the Chalais business in 1626 partly to support Marie's favorite son, Gaston d'Orléans, and Richelieu had subsequently demonstrated his ability to exact cold revenge. Some of Richelieu's former supporters began to pull away, such as one of his chief propagandists, François Dorval-Langlois, sieur de Fancan. Fancan began to write against the Jesuits and was found to be in correspondence with powers in England and The Netherlands, as well as with Ferdinand, archbishop of Cologne (brother of the Duke of Bavaria). In September 1627—at the time of the siege of La Rochelle, not long before Descartes's own involvement—Fancan was imprisoned in the Bastille; he would die there in 1628, accused of advocating a "republic" (of the great nobles?).[84] But he was only one example of people caught up in Richelieu's efforts to suppress all hint of opposition. At the end of February 1629, Alexandre de Vendôme, a son of Henri IV and Gabrielle d'Estrées, died in the prison of Vincennes, where he had been kept following the Chalais conspiracy. In the meantime, Gaston d'Orléans would flee to Lorraine, where he and the duchesse de Chevreuse, together with the ducs de Guise, plotted with the queen mother against king and cardinal. The opposition also included the duc de Bellegarde, once a client of d'Epernon, with whom Balzac was associated.

The growing differences between the queen mother and Richelieu also brought into play Cardinal Bérulle. Richelieu seems to have tried to move the rival cardinal aside, for Bérulle was offered the archbishopric of Tours in October 1628 (shortly before the moment when Baillet dates the meeting with Chandoux); had he taken it, the position would have caused him to turn much of his attention away from Paris. But Bérulle turned down the offer—a rare instance of a courtier declining a lucrative honor—to keep his hand in the game at court.

The next year would see Bérulle's mysterious death. Richelieu had persuaded the king to take quick action in support of Mantua by leading his army into Italy in person, even though it was the winter of 1628–29. The queen mother was left in charge of the government during her son's absence, and Bérulle sat on her council. The differences between the two cardinals were public by that time (for instance, Richelieu wished to renew the treaty with the United Provinces while Bérulle was opposed). But in the field, king and cardinal achieved victory at Casale and followed it up with a campaign against the Huguenot cities of the south of France, removing any further chance of Spanish intervention there. As the king's army moved through the region, it sometimes committed atrocities, as at Privas, preventing the Huguenot leader, the Duke of Rohan, from finding a secure place to make a stand. Aristocrats such as Béthune, on whose behalf Descartes had acted previously, took Marie de Medici's view that Richelieu was destroying France internally while bringing its interests into danger abroad, and withdrew from government. The Edict of Alais, signed at the end of June, finally put an end to the political privileges of Huguenots; while the king returned to Paris, Richelieu continued to head the armed campaign against the remaining Huguenot cities, and then he turned again to Italy. When Richelieu finally returned from the victorious campaigns, Marie de Medici received her former adviser not with warm celebration but with haughty coldness. Descartes himself later expressed strong aversion to the bigoted and superstitious zealots who "imagine they are such close friends of God that they could not do anything to displease him."[85] When Bérulle died suddenly at the beginning of October 1629, "everyone" knew that he had been poisoned by Richelieu.[86]

At the end of 1628, then, Italy was getting hot, and the heat would soon be reflected back on Paris, touching people known to Descartes. As papal envoy, Bagno had to watch every move carefully, particularly with regard to events in Italy, where Rome's interest was to achieve a negotiated settlement. The warmth of the moment raises the possibility that Descartes may not have been called out by Bérulle so much as called on by him—especially if the talk was given by someone associated with policies advanced by Richelieu, such as Chandoux, who might be open to criticism.

Consequently, it is possible to imagine the following scenario: once the notables had returned to Paris after the siege of La Rochelle, the papal nuncio hosted a gathering to hear from Chandoux as a proxy for the great persons he was cultivating. Someone of similar views, Renaudot, was setting up the Bureau d'Adresse under the protection of Richelieu. The bureau was planned to be concerned not only with new ideas but also with good works, a grand ambition that also attracted the interest of Bérulle and other clerics, all concerned about the poor. Neither Richelieu nor Renaudot is mentioned as attending the *assemblée*, as they were not directly concerned, but everyone could be certain that what was said would be reported onward, so Chandoux could serve to take the temperature of the moment. Perhaps Chandoux was being held out by Bagno as the cardinal's proxy. A minor but valorous nobleman with sharp skills at methodological criticism, known to both the host and Bérulle, was also invited. Perhaps Bérulle used Descartes as a stalking horse, publicly asking him to comment when knowing he would attack. In any case, Descartes well understood the importance of the moment. He first deferred to others, and then only criticized a part of what was said after being asked directly by a senior member of the government for his opinion. But he would have been identified as Bérulle's client. Whether intended or not, then, the outcome of the meeting could easily have been read as signaling opposition to the reforms being promoted by Richelieu. Although afterward being embraced by Bérulle, Descartes had ventured his fortune on the losing side. He had been outed as a member of a "sect."

Feeling Threatened

Whatever one thinks of such a speculative line, it is clear enough from other evidence that Descartes was at growing odds with Cardinal Richelieu's regime just when others were, as well, and he had found the courage to speak up. Everyone knew that danger lurked in any position that was not simply and completely loyal to the cardinal and his projects.

Baillet comments that after leaving France, during his first years in The Netherlands, Descartes did everything possible to keep himself concealed, even resorting to having letters and packages meant for him sent to friends, from whom he could retrieve them.[87] But the best evidence that Descartes felt threatened comes from his good friend Balzac. Within a few years, Balzac would do his best to make peace with Richelieu, although he spent most of the rest of his life living away from Paris. But apparently Descartes would not try accommodation, refusing to visit France until after the cardinal's death in 1642. Perhaps he never forgave Bérulle's death—which deeply troubled him[88]—for in a letter to Balzac to which we will turn to in a moment, he refers to "assassinations."

Balzac, we know, was feeling vulnerable in the later 1620s. His former friend and fellow traveler, Théophile de Viau, had been imprisoned and tried as an example to libertines and former libertines, and some of the same people who set up Théophile were now after Balzac. Balzac had once personally served Archbishop La Valette (of d'Épernon's family), but from 1623 onward Balzac's relationship with La Valette decayed, leaving him not only without a major protector but with a potential enemy, since La Valette would become one of Richelieu's allies. Balzac's literary ambitions were in keeping with the project of his generation, to make French prose as supple as French poetry, allowing it to reign over Latin and Greek, and Balzac became known as a stylist for his skill with letters. He frequented the salons of Marie le Jars de Gournay and Catherine de Vivonne, marquise de Rambouillet, and his collection of *Lettres* (1624) had established his public reputation. But he also had severe critics, who accused him of "vanity, veiled obscenity, and mendacious use of hyperbole," code words for dissimulation and libertinism.[89] Cowed, at the time of Théophile's arrest Balzac renounced all dangerous ideas.[90] Descartes tried to help his friend by traveling to Fontainebleau to recommend Balzac to the visiting Cardinal Barberini. But in the spring of 1627, Balzac published a work under the pseudonym of "Augier" to restore his reputation by praising himself, and his ruse was uncovered. The incident served only to raise additional questions about his personal integrity. Worse, he had dedicated the work to Richelieu, who was embarrassed and listened to complaints against the author, including a charge of plagiarizing from ancient writers. Together with a friend, the prior François Ogier, Balzac managed to rebut the accusation of plagiarism, but his counterattack on his enemies was choleric enough to work against him again, bringing him to the attention of one of the major antilibertine figures of the period, the Feuillant theologian Jean Goulu.[91] It was a dangerous moment for him.

It was about then that Descartes wrote a highly favorable Latin appraisal of Balzac's work. No one knows exactly when it was written or quite for whom it was intended, but everyone would have known that it would end up on Richelieu's table. It is likely to have been written early in 1628, since on March 30, 1628, Balzac wrote to Descartes with copies of three of his "Discours" dedicated to Descartes—an implied sincere thanks—which would later appear in his posthumous *Socrate chrétien* (1652), and asking for Descartes's account of his progress in finding a method of attack against the "giants" of the schools, which he and his other friends hoped to see soon.[92] Descartes's defense of his friend was a spirited plea for what we today would call free speech: he not only wrote in praise of Balzac's rhetorical skills but also stated that if his friend ever "undertakes to depict the vices of the Mighty, he is not prevented from speaking the truth by a servile fear of power."[93] Descartes was in effect acting as his friend's champion, daring his enemies to take him on.[94]

Cardinal Richelieu found a sinister avenue of attack on Balzac. Balzac and another acquaintance of Gournay, Honorat de Bueil, seigneur de Racan (a noted poet, and soldier), got into a private quarrel in which Balzac accused Racan of being eaten away by syphilis to the extent that he had become incapable of paying proper homage to women. Racan replied in kind. Their letters were intercepted by Richelieu's spies and published, shaming them publicly. At the end of 1628, his personal reputation in tatters, Balzac retreated to his country house, where he remained in a kind of internal exile for three years.[95] About the same time, the salon of Madame des Loges, which both Gaston d'Orléans and Balzac had attended, was closed by Richelieu. Gaston retreated to Blois; he considered leading the army into Italy, but upon hearing of the death of his dear half brother, Alexandre de Vendôme, in the prison of Vincennes, he turned around. By July, he had fled to the principality of Lorraine, which Richelieu would come to destroy. When Alexandre's brother, César, was released from prison at the end of December 1630, he fled to Holland to fight beside the Prince of Orange, remaining an anti-Richelieu exile abroad until after the cardinal's death.[96] A cold threat was spreading, with assassination and even execution becoming real ends. Retreat became common.

Exile

Descartes left France entirely. Beeckman saw him again in Dordrecht at the beginning of February, and he starts to appear in other sources from the Dutch Republic thereafter. Only in the summer of 1644, after Richelieu's death at the end of 1642 and Louis XIII's in May 1643, would Descartes make his first brief return. As we have noticed him doing before, however, on his departure in the winter of 1628–29, Descartes took suitable precautions to put people off his scent: he left a trunk with his cousin "to make people believe he would soon return."[97] A year later, in April 1630, he dropped a comment in a letter to Mersenne about composing a treatise "if I am still living." Go ahead and spread his ideas, he said, but please "do not mention my name."[98] He seems to have felt fear to his bones, and to have worried about putting others at risk, too.

In his *Discours* of 1637, Descartes wrote briefly about the move. He had until then refused to take sides "regarding the questions which are commonly debated among the learned." But he found that some people were circulating the rumor that he had found a foundational certainty for the new philosophy, which would make him visible. "I cannot say what basis they had for this opinion," he wrote. We might think of his defense of Balzac, but that was a private matter. Presumably, then, the debate at the papal nuncio's brought him to the attention of the Parisian learned world, for "I confessed my ignorance more ingenuously than is customary for those with a little learning, and perhaps also because I displayed the reasons I had for doubting many things which others regard as certain." But now his friends urged him to stand up, and "I was honest enough" to try to live up to their expectations. Therefore, "exactly eight years ago this desire made me resolve to move away from any place where I might have acquaintances and retire to this country," the Dutch Republic.[99] Since his book appeared early in 1637, he points to early 1629, which fits with what else we know.

Descartes implies that by standing up for his friends he would have been putting them in danger, which is supported by extant correspondence. In a letter of April 15, 1631, to Balzac, Descartes said he had wanted to write much earlier but had restrained himself, "though you owe me nothing on that score." He had not wanted his friend to receive a letter from him until Balzac had been accepted back into Parisian society—Descartes's own personal reputation seems to have been more a problem than an asset for Balzac.[100] In the *Discours* he added that the many decades of warfare in the low countries, with a relatively stable military border south and east of the places he resided, "has led to the establishment of such order" that "the fruits of peace" were entirely secure. He could therefore live among "this great mass of busy people," who were entirely concerned with their own affairs, "as solitary and withdrawn as if I were in the most remote desert, while lacking none of the comforts found in the most populous cities."[101] In other words, he could live at ease in the Dutch Republic without worrying about opinion in Paris, or threat.

Let us unpack the moment of his leaving. Descartes had many friends in Paris, some of them important, several of whom were becoming associated with discontent. He thought he could not afford to be linked to any sect, or party. He had gone separate ways from his father due to the Chalais trial. He had recently participated in the campaign for La Rochelle, but he did not join the king's army in the Italian campaign, perhaps seeing events in a way aligned with the queen mother. He had been in The Netherlands before and found it to his liking. It was again engaged in a bitter struggle with Spain, but the fighting was far from the province of Holland. The Dutch Republic remained a French ally, but safely distanced from Richelieu's direct intervention. There would be plenty of opportunities to invest any remaining funds he had acquired in Italy. He headed north.

That he had serious concerns about what France had become under Richelieu is further confirmed in his later letters to Balzac. After their initial separation, the two friends communicated only with great caution—letters could be read by others, too. The letter to Balzac of April 1631 indicates that Descartes had had a (now missing) message from

Balzac eighteen months earlier—which would place it around November 1629—reporting that he hoped to return to court. But apparently he could not, since in 1631 Descartes expressed his sadness that his friend "had to stay there," at his country house, "as long as you have."[102] Now that Balzac had finally been welcomed back to the capital, though, Descartes felt free to reply at last.

By 1631 Balzac believed that he had finally made peace with Richelieu. He had just finished a book (*Le Prince*) in which he praised the cardinal for restoring France to health after a long period of decrepitude. It was one of the most ardent defenses of the cardinal's policies published in the period. "Beneath the same faces I see different men, and in the same kingdom, another state. The outward appearance remains, but the interior has been renewed. There has been a moral revolution (*une révolution morale*), a transformation of spirit," he wrote.[103] In March 1631 he presented an advance copy of the book to Richelieu himself, who declared that he was satisfied. Balzac returned to Paris shortly thereafter. Descartes now felt able to break his silence, and even to think of returning himself. "I must say that during the two years I have been away I have not once been tempted to return, until I was told that you were there." It only now occurred to him that he could be "happier somewhere other than where I am now."[104]

But apparently Descartes's personal fate would not yet allow him to step out of the shadows. He was deep into a mysterious task "which keeps me here . . . the most important one I could ever devote myself to." He adds, "Please do not ask me what this task that I deem so important might be, for it would embarrass me to tell you." Is he simply expressing false modesty about writing philosophy? Galileo had not yet been condemned, so why hide such a thing? Or was Descartes still in contact with the house of Lorraine, or the former queen regent or her son, Gaston, or others among the nobles or clergy who were opposed to Richelieu, at the time when Balzac had become a propagandist for the cardinal? Or was he simply expressing dissimulation in a manner that Balzac would have appreciated, having now "become so philosophical" that he "despised" things that were "ordinarily valued" and valued things that were ordinarily "put at no value"? His world was upside down compared to Balzac's. But he had found a path ahead. "I shall be content to tell you that I am no longer of a mind to commit nothing to paper, which as you saw, was once my intention." He was only now beginning to feel able to use his quill again. But he remained elusive, apparently fearful of putting anything in writing even to a dear friend. "I shall tell you about it more openly someday if you would like me to."[105]

Descartes's letter goes on to say that the main problem in considering a return to Paris was his own reputation. Compared to Balzac's, he had only an "indifferent and uncertain" status, which would not protect him in Paris. Even with a dear friend at his side, Descartes still believed his reputation would not be able to sustain the dangers there. But in Amsterdam he had peace of mind. "Here I sleep for ten hours every night, and with never a care to wake me," before rising to mingle his daydreams and night dreams, allowing his senses to share in all the allowable pleasures of life. In brief, all he missed was Balzac's conversation, the pleasures of which almost tempted him back to Paris, where he could have told him in person that he was, "with all my heart, your very humble and devoted servant."[106] But he balked at a personal return. Balzac was now cozying up to the cardinal with all the flattery at his command; might his invitation be a ruse?

Balzac must have replied straightaway. Unfortunately, he had not conquered Paris after all. Despite Richelieu's personal acceptance of his book, the conservatives had read it carefully and continued to question Balzac. They even asked the Sorbonne to look into the orthodoxy of some passages—which would cause the theologians to express concern in December—and at some point Balzac retreated to Blois and begged the cardinal to show a public sign of his favor. The minister remained cool, however. Apparently on further reflection, despite honeyed praises thrown his way, Richelieu had taken offense at Balzac openly mentioning in print the difficulties of his relationship with the queen mother: after the "Day of the Dupes" on November 11, 1630—when she seemed to have secured her son's agreement to rid himself of the cardinal but he in turn persuaded Louis to let him stay, allowing him to turn on all who had celebrated his only apparent fall from grace—she had fled north to Compiègne, and then to Brussels, a political satellite of Madrid. In Paris, the worst of the anger about Balzac's fawning book would blow over, but in Brussels *Le Prince* would be burned (presumably for its praises of the cardinal).[107] Balzac eventually took a seat in the Académie française after its foundation in 1635 (as number 27),[108] and he published a new edition of his well-regarded *Lettres* in 1647. But he would spend most of the rest of his life in Angoulême: he had problems with his health, it was said.

Having heard of Balzac's disappointment, then, in a return letter of May 5, Descartes expressed his surprise that Balzac was contemplating joining him in Amsterdam. He had imagined Balzac reentering Paris in triumph. But he commiserates with his friend about the news that "a mind as great and generous as your own should not be able to adapt itself to the constraints of service to which one is subject at Court." The "constraints" were great enough to induce Balzac to consider retiring from "the world" (i.e., France) entirely. Tongue in cheek, Descartes went on to praise Amsterdam as a far better retreat than any Franciscan or Carthusian monastery, or even "the famous Hermitage" where Balzac had spent the last year (Blois?). Even in a country house one has visitors, he noted, but in Amsterdam "everyone but myself is engaged in trade, and hence is so attentive to his own profit that I could live here all my life without ever being noticed by a soul." He could walk among the bustling crowds as if it were a leafy grove empty of other humans, noticing the workers as Balzac would notice rustic peasants in the countryside, whose activities not only serve to provide all one's needs but "serve to enhance the beauty of the place." In Amsterdam, too, one could find "all the conveniences of life and all the curiosities you could hope to see" brought there from all the Indies and Europe.[109]

But then, hinting at darker motives, he asks, "In what other country could you find such complete freedom, or sleep with less anxiety, or find armies at the ready to protect you, or find fewer poisonings, or [fewer] acts of treason or slander? Where else do you still find the innocence of a bygone age?"[110] It was far preferable to the dangers of Italy, he went on (seeming to imply the use of Machiavelli's methods in the country where Balzac still resided). In short, for Descartes, the peace and quiet of The Netherlands meant pleasure and, above all, personal safety. No poisoner of reputation or life awaited him. He would be happy to welcome his dear friend to a place where he would not have to fear Cardinal Richelieu.

PART 6

Not Yet Concluded

We promised ourselves to watch the youthful sieur du Perron closely as he moved about in the foothills. Now, early in 1629, René Descartes was in The Netherlands; in March he would turn thirty-three. His friendship with Guez de Balzac and others, and his efforts to rise among the aristocratic circles of the kingdom through war and diplomacy hint at the passions that moved him. We have found him traveling throughout most of the European continent—only Spain excepted—often in the presence of the great powers of the day. He moved far and wide, several times returning to Paris, but now he was off again, again not knowing what would happen to him next. It would be from The Netherlands that he would decide to tackle the mountain at last, from there that he would publish his books. He had acquired the latest techniques for climbing, and the best advice. He may not have set out to become a mountaineer, and he had watched many others attempt it and fail, often at the cost of their reputations, even their lives. But by now he was as courageous and accomplished as any. Looking up from the low lands, he sensed a route that could get him safely to the top and down again if he moved with caution and determination. His friends had urged him on and offered help. He had bet on hope before, and won. But it was not an ambition for a settled life, and he would shift his living quarters many times more even within the orbit of the Dutch Republic, making several further trips elsewhere, too, before meeting his unexpected end on yet another journey, well before old age, in Stockholm.

Descartes's departure from Paris has meant that he came to be seen as a kind of loner in search of the quiet life. But he had many friends and had long been engaged in some of the chief events of his day. He showed a deep appreciation for the need for cooperation—in private life, in public life, and even in pursuit of the new science. "By building on the work of our predecessors and combining the lives and labors of many, we might make much greater progress working together than anyone could make on his own," he wrote.[1] Descartes's written plans for the establishment of a Swedish academy of sciences on behalf of Queen Christina are sometimes noted. The informal *assemblée* (assembly) of natural philosophers that he and Nicolas Le Vasseur hosted has, however, disappeared from the history books: aside from Baillet's brief listing of its members, it is absent from later accounts of his life and work, causing him to seem to be simply one of Marin Mersenne's distant correspondents. An invitation from Charles Cavendish to Descartes to relocate to England in 1640 and begin a scientific academy there, reported by Baillet, has often been overlooked, too.[2] Descartes would retain links to Parisian debates via correspondence and a few later visits, but not in person. Looking outward from France toward the "philosopher" Descartes, then, made him seem far away and on his own, even when in The Netherlands he established good relationships with Dutch intellectuals and politicians as well as French officials. He has ever after been represented as a kind of monk, inventing new worlds of the mind in the privacy of his study.

Instead, we have seen Descartes as a figure engaged in the life around him rather than isolated from it. Many of his battles appear to have been waged out of sight, however: his slipping away for days at a time while still in Paris, the effort to throw people off the scent when he left for the Dutch Republic, the enormous trouble he first took in corresponding with Mersenne via intermediaries so that people in Paris would not know his address, the important project on which he was engaged but could not confide to Balzac, the distinction he made between private and secret papers when he left The Netherlands for Stockholm. What he was up to in Germany and Italy is also unclear, as were his travels at the edge of the Baltic. We will probably never know the full range of his actions or intentions. His behavior certainly gives him an aura of reticence, even sometimes of deliberately hiding his movements. But although further investigation may show otherwise, at the moment there is no evidence that he was in the employ of a state, an aristocratic house, or a religious organization. It is probably best to assume that his engagements arose from trying to find his way into the patronage networks of his time and not quite succeeding, although the possibility that he was gathering intelligence for someone, or someones, must remain open.

Descartes came from a Catholic and loyalist family of *politiques*—that is, people who were eager to have strong but lawful governance, accepting a variety of customs and beliefs as long as civic order was upheld. His own political and cultural sympathies seem to have aligned with the government of his youth, that of the French king Henri IV and his queen, Marie de Medici, which aimed toward inclusive and outward looking, nondoctrinaire but aristocratic, charismatic monarchy. All the evidence points to his remaining loyal to Her Majesty's friends in years after. When the autocratic reign of the royal couple's son Louis XIII announced itself after the assassination of Henri IV, Descartes's path began to diverge from the new king's. The closest friend we know of from Descartes's youth is Guez de Balzac, whose family served Jean Louis de Nogaret de La Valette, duc d'Épernon, Marie's chief support in the wake of her son's violent coup. Even before then, Baillet tells us, Descartes looked for favor to the cadet branch of the house of Lorraine, most likely through one of Marie de Medici's favorites—also loved by Louis—the duc de Chevreuse, who had powerful personal interests elsewhere in Europe as well. If that is so, in the mid-1620s Descartes's interests would also have overlapped with one of the chief members of the discontented aristocrats, Marie de Rohan, who became the duchesse de Chevreuse and took the side of Marie and Anne of Austria in contests with the king. Several of Descartes's associates, including Mersenne, were in circles near Louis XIII's younger brother Gaston d'Orléans, the perpetual heir in waiting and ally of the House of Lorraine. Cardinal Pierre de Bérulle, with whom he had a private audience, remained one of Marie's closest advisers; his landlord, Le Vasseur, served the great Louis de Marillac, another of Marie's favorites.

But by the later 1620s, all such people, including Balzac, were moving further away from the king and his chief minister, Richelieu, who had begun as a client of Marie de Medici but shifted his interests to the king, further dividing mother and son and their parties. In the 1630s, Descartes would have watched from exile as Cardinal Bérulle died suddenly, Marie de Medici fled, Gaston d'Orléans was sidelined, Marie de Rohan went into exile, and Lorraine was invaded. After the deaths of Louis XIII and Richelieu, however, Descartes would seek the patronage of the government of the survivor, the queen regent Anne of Austria, although yet further factional conflict meant he would never settle again in his beloved Paris. As far as we can tell, then, in leaving France he seems to have remained loyal to the old ideals while avoiding the continuing political sectarianism of the realm—although he kept in close touch with French ambassadors and possibly others as well.

Perhaps at first he was also renewing his contacts with Dutch military engineers? Descartes showed up in Dordrecht at the beginning of February 1629 to visit his friend Isaac Beeckman, but in April he matriculated at the university in Franeker, from where he is known to have addressed four letters through September. There the famous instructor on mathematics and military engineering, Adrien Metius, was lecturing. A month later, the Prince of Orange began the great siege of s'Hertogenbosch (which surrendered in September). Curiously, the savant Pierre Gassendi visited the scene of action in person. Gassendi had also left Paris about the same time as Descartes (around Christmas 1628), and although he spent most of his time in the Spanish Netherlands (where he met, among others, the chemically adept physician Jan van Helmont) he visited the Dutch Republic briefly in July. When he did, he met twice with Beeckman—with whom he had not previously been in contact—and then visited the siege works. His movements suggest advice from Descartes, although like so much else, that suggestion must remain speculative. When Gassendi returned to Paris, he was excited about the possibilities of Epicurean natural philosophy, on which he would write in coming years. Within two weeks of Gassendi's return, Mersenne was planning his own visit to the low countries, which he undertook in 1630, meeting Descartes in person.[3] About the time Mersenne set out, in April 1630, Descartes wrote to him about how he had begun a project "to explain all the phenomena of nature" and that "I am now studying chemistry and anatomy simultaneously; every day I learn something that I cannot find in any book." He hoped to have it completed within three years, "if I am still living." But he did not want it known, so that he would "always be free to disavow it."[4] As he later explained to Balzac, he was involved in something he would not explain on paper. Unlike Gassendi and Mersenne, he would not return to France.

Moreover, in the years before the condemnation of Galileo in 1633, although Descartes was never simply a conforming Catholic, he would remain on good terms with members of the papal court. He was not only acquainted with the Oratorian Cardinal Bérulle but also knew the pope's nephew, Cardinal Francesco Barberini, well enough to ask for protection for his friend Balzac; he spoke against Nicholas de Villiers, sieur de Chandoux, at a meeting hosted by the papal nuncio, Giovanni Francesco Guidi di Bagno, one of Pope Urban VIII's closest confidants. And in 1632–33 he would be recommended for a professorship at the papal university in Bologna by yet another cardinal, nuncio Francesco Adriano Ceva, not only a trusted adviser from Urban's youth but rumored to have been the person who brokered the deal that put him on the throne of Saint Peter in 1623.[5]

Ceva recommended Descartes for the chair in theoretical medicine shortly after Urban VIII had personally approved the publication of Galileo's *Dialogue on the Two Chief World Systems*. At the time, the pope was rumored to look favorably on Gustavus Adolphus, who with his decisive military victories was shaking the foundations of Habsburg rule in central Europe. In France at that moment, Richelieu had executed Marie de Medici's loyalist, Marillac, and an aristocratic uprising was being led by Gaston d'Orléans and the duc de Montmorency, marshal of France (and former protector of Théophile). Richelieu assumed that if the rebels had some early victories, the king of Spain and the duc du Lorraine would come to their aid with additional troops, and a general rising against the king might have been successful.[6] On September 1, however, the outnumbered royal forces won a battle at Castelnaudary, in Languedoc, resulting in the capture of the wounded Montmorency and his abandonment by Gaston; Montmorency himself was executed shortly thereafter.[7] A few weeks later the king of Sweden met his end at Lützen, and Galileo would be ordered to travel to Rome to face the Inquisition. When Descartes heard the news of Galileo's formal condemnation in the summer of 1633, he abandoned work on his account of the universe, *Le Monde* (*The World*), and with it the medical treatise he was writing as one of its parts. He would never publish those works during his lifetime.

But to all appearances, he lived a strangely masterless life. Not that he would have been looking for employment: a person of Descartes's rank would be trying to please those he honored and freely accepting rewards from them, but spurning fees for service. Although he once accepted payment for soldiering, that seems to have been the extent of it. Late in life he became one of the philosophers of the queen of Sweden, but that was a position of favor rather than office. While Descartes had once contemplated taking a position as a government official in France, he never did so; he seems to have had financial interests but not commercial ventures; as he later replied to his critics, he certainly never took religious orders, much less a vow of chastity. One of the reasons given for his extremely humble burial was not to go to expenses that would be have to be passed on to his family:[8] he seems to have lived relatively modestly, on his own income or credit. One might wonder what his future would have been like if he had taken up residence in Châtellerault or Poitiers, putting his shoulder to the wheel in service to the city or kingdom and raising a family. Or what might have been, had the comte du Bucquoy not died in battle, or had Descartes stuck with the Duke of Bavaria or Philippe de Béthune, or acquired the office of intendant of the army in the Piedmont, or if Princess Elizabeth of Bohemia joined one of the princely houses of Europe in marriage and brought along her adviser instead of becoming a powerful abbess in Herford. In any case, it seems his initial hopes had been foreclosed by the early "accident" (as Baillet called it) of November 1617, the bloody coup of Louis XIII, which set events in motion that would lead to the rise of Richelieu and the disaffection of many of the great nobles who remained loyal to Marie de Medici, with whom Descartes would be most closely associated.

His time in the Dutch Republic would not be easy, either. On the one hand, the new Prince of Orange, Frederick Hendrik, did not depend as much on religious militants to keep order as had Mauritz, and so after 1625 the Remonstrants began returning from exile and resumed places of importance in such cities as Amsterdam. Jews were tolerated in several cities, and even Catholics—who composed a large proportion of the population—were usually allowed to worship as long as their churches were not visible from the street. The Dutch Revolt had been fought in the name of freedom of conscience. But on the other hand, while private views were for the most part allowed, anything threatening to undermine public order could lead to trouble. When Descartes told Balzac that he had found a place without poisonings, slanders, or treasons, then, we should read the comment comparatively rather than absolutely.

A fresh study of his two decades in the Dutch Republic would show much of interest, but it was clearly not simply a place of quiet for Descartes. An example of the sensitivities that could make life dangerous for outspoken nonconformists is the indictment and trial of Johannes Symonsz van der Beeck, better known by his Latinized name, Torrentius. Its denouement was wrapping up just when Descartes arrived. It echoed the fears in Paris about a Rosicrucian takeover, which, as in Paris, lumped together the Rosicrucians and the libertines, with Torrentius becoming a scapegoat as Théophile de Viau had been. Torrentius had become one of the most skillful representational painters of his generation—or a very accomplished user of the camera obscura—although because so many of his paintings were sexually explicit, only one of them, a still life, is known to have survived. Born into a Catholic family, he gained a reputation as a moral and philosophical libertine, being identified in the early 1620s by church authorities in Haarlem as a notorious heretic and jailed for not being at one with his wife. When on January 29, 1624, a few months after the Rosicrucian fright in Paris, a letter was sent from the States of Holland and West Friesland to the Court of Holland warning about a sect called "La Roze Croix" that was now in Holland and particularly strong in Haarlem, Torrentius came under even more serious scrutiny. After long investigation, Torrentius was identified as one of the principal threats to morals, and a special legal process was begun against him that allowed no defense to be mounted; in late 1627 and early 1628, he was interrogated and on several occasions put to severe torture, but he refused to confess. Despite support from the Prince of Orange, his legal appeal was turned down, and for leading a scandalous and blasphemous life (without further reference to the Rosicrucians) the prosecutor asked that he be burned at the stake; instead he was condemned to twenty years in prison. A few months later his associates were banished from the city as well. His influence was even blamed for the mutiny and wreck of the East Indies ship *Batavia* off the coast of Australia in 1629, with its subsequent rapes and murders. In 1630 Torrentius was granted a pardon by Frederick Hendrik to flee to the English court in London, only to return when civil war broke out there.[9]

Descartes would have to continue to watch his step. One of Beeckman's students in Rotterdam had a brother who had lodged with Torrentius in 1626, shortly before the legal process was begun against him; and Descartes's good friend Cornelis van Hogelande, a physician, chemist, and reputed Rosicrucian, belonged to the same Amsterdam dueling academy as Torrentius, which was run by Gérard Thibault (fig. 14).[10]

Figure 14. Thibault's Method of fencing geometrically. Reprinted from Girard Thibault, *Academie de l'Espée de Girard Thibault d'Anvers* (Antwerp, 1628). Courtesy of Anne S. K. Brown Military Collection, Hay Library, Brown University.

One of Torrentius's lewd paintings had a phrase from Ovid inscribed on it—*qui bene latuit bene vixit* ("He who lives hidden, lives well")—which also appears on the portrait of Descartes by Edelinck.[11] Not surprisingly, then, when in the later 1630s Descartes's views came under public scrutiny because of the conflict between the medical and theological faculties at the university in Utrecht, a spokesperson for the opposition, Martinus Schoock, accused him of being another Vanini or Torrentius.

Descartes had heard such dangerous accusations by association hurled at others years before, in Paris, and he remained vulnerable. As a private person without formal institutional membership, even more as a foreigner, he clearly felt the threat. His *Meditations* (1641) publicly established his belief in God, which had been questioned.[12] He clearly developed strong resentments, explaining privately to Princess Elizabeth that the scholastic theologians were ganging up to try to "crush" him "by their slanders," and telling a student who visited him in April 1648 that he was very sorry scholastic theology had not yet been "stamped out," warning: "We must never allow ourselves the indulgence of trying to subject the nature and operations of God to our reasoning."[13] With the help of his friend Henri Brasset, the resident for the French embassy in The Hague, Descartes obtained the support of the ambassador, Gaspard Coignet de la Thuillerie, as well as the Prince of Orange, and responded to the libels and court proceedings as vigorously as the law allowed.[14] For a time, he must have been constantly occupied with reading and drafting letters and legal positions. Peace and quiet is not what his life in The Netherlands would amount to after all.

There were attempts at reconciliation with France as well, further suggesting that Descartes had left because of a problem with Richelieu. After the deaths of king and cardinal, in the summer of 1644, Descartes visited Paris again, residing with his good friend, Abbé Claude Picot. He also traveled to Brittany for a couple of weeks in July, presumably to touch base with his relatives, bringing with him a draft of a translation into French of his *Meditations*. The translation had been begun by Louis-Charles d'Albert, duc de Luynes, eldest son of Marie de Rohan, who had herself returned from exile to resume her place among the group of discontented nobles of France known as the *Importants* before being sent away to her country estate by Queen Anne. Her son was serving in the wars against the Habsburgs as a staff officer to the duc d'Enghien, later known as the Grande Condé; d'Enghien and de Luynes were themselves under the nominal command of Gaston d'Orléans, who was proving himself a successful general. For the moment they must all have been breathing more freely. With help from Descartes himself, De Luynes's translation would be published in 1647.

Descartes also made new acquaintances in Paris, leading to a royal pension. He met English virtuosi exiled from the civil wars, including Kenelm Digby and the Cavendish brothers Charles and William—soldier savants like himself—and, more important for his personal fortunes, Claude Clerselier, a member of the *parlement* who became one of his chief advocates, the first editor of Descartes's letters and posthumous works as well as the initiator of the biographical study later published by Adrien Baillet. Clerselier introduced Descartes to his brother-in-law, Pierre Chanut, an official who became the most senior French diplomat in Sweden. Chanut and Descartes also felt a common understanding, no doubt in part due to their years living abroad. The diplomat in turn introduced him to the queen regent's chancellor, Pierre Séguier. (After the death of Louis XIII, Séguier had overridden the king's wishes and rallied support for Anne's regency in the *parlement* of Paris.) Chanut and other friends started advocating a royal pension for the philosopher. But for some reason Descartes turned around again. By mid-November 1644, he and Picot had returned to the Dutch Republic, where he would settle for the next three years in Egmond-Binnen.[15]

In The Netherlands Descartes resumed his role as adviser to the Princess Elizabeth. He had been introduced to her in 1642 by Alfonso Polloti (or Alphonse Pollot), a Piedmontese general in Dutch service. But in 1646 Elizabeth was sent away to relatives in Brandenburg for her staunch defense of her brother, Philip: he had killed a French colonel, Jacques de l'Epinay, in a duel, after L'Epinay publicly boasted of sleeping with both Elizabeth's sister, Louise, and her mother, the Winter Queen of Bohemia. (Philip would later take service with the Duke of Lorraine, only to be killed in the uprising of the Fronde.)

With Elizabeth gone, De Luynes's version of Descartes's *Meditations* nearing completion, and the matter of the royal pension coming to a conclusion, Descartes returned to Paris yet again, in June 1647. He took up residence in the rue Geoffrey-l'ânier, in the same house as the family of the young Françoise d'Aubigné—later famous as Louis XIV's Madame de Maintenon—whose family had also been in opposition to Richelieu.[16] By now, Descartes had become a well-known author: following the condemnation of Galileo, Descartes's earliest work had ended in self-suppression, but the *Discours* and its accompanying *Essais* had been published anonymously in French in 1637, and "everyone" soon knew him to be the author; he had published his *Principles*, for the savants, in Latin, in 1644; the Latin *Meditationes* was now about to appear in accessible French. He had also begun working on a study that would be published in 1649, in French, as the *Les Passions de l'Âme* (*The Passions of the Soul*).

Descartes visited Brittany on family business again in late July, but in Paris he continued to make new acquaintances. On September 23 and 24, for instance, he visited the young Blaise Pascal, who had recently published on new experiments that proved the existence of the vacuum, a position Descartes had long argued against. (A year later, Pascal's brother-in-law would carry out the famous Puy-de-Dôme experiment that confirmed the findings.) There were sad disappointments, too, however: Descartes's old friend Mydorge died in July, and Mersenne was seriously ill.

By late in the summer of 1647, all was finally in order to receive a pension from the chief minister, Mazarin, in the name of the boy king Louis XIV. The formal document awarding him three thousand livres per annum was handed to Descartes in Paris on September 6. It was granted in consideration for the "great merits and usefulness that his philosophy and investigations (*les recherches*), after long studies, have procured for humankind; as also to help with the continuation of his wonderful experiments (*belles experiences*), which require expenditure."[17] Indeed, the costs of his earlier work on optics and his continuing anatomical researches into animal bodies must have been substantial. His studies had been formally recognized as useful and *belles*. Given Descartes's later complaints about the cost of worthless parchment, however, it is unlikely that he ever collected any income from it.[18]

Descartes then left Paris yet again. Picot once more accompanied him to Egmond and remained with him until January.[19] Perhaps Descartes had unfinished business in Holland, where a peace treaty was being negotiated that would end the Eighty Years' War fought against Spain but would also take the French ally out of the war against the Spanish Habsburgs. The French resident in The Hague, Henri Brasset, shared with Descartes the hope that the young new stadholder, William II, Prince of Orange (married to the Anglo-French Mary Henrietta, daughter of Henrietta Maria) would continue to help France—that hope would only be dashed by William's death from smallpox in 1650.[20] The Peace of Westphalia would end the war in the Holy Roman Empire but also begin to unravel the alliance between France and Sweden that had finally won the day; from Sweden Ambassador Chanut began a serious effort to lure Descartes to Stockholm to become an adviser to the queen. Diplomatic entanglements continued to bind.

First, however, Descartes returned to Paris again, although again he did not stay. He arrived in the spring of 1648, just in time to find the *parlement*—where he had several friends—meeting every day to discuss the tax that Mazarin had imposed on them to keep the war effort against Spain alive. He was clearly keeping a close watch on events. He also visited and reconciled with Gassendi and for the first time met Antoine Arnauld, a leading member of the rapidly growing dissenting Catholic group known as the Jansenists, who were also strongly supported by Marie de Rohan and the duc de Luynes.[21] At the end of August, however, following Condé's great victory over the Spaniards at Lens, Mazarin finally lost patience and arrested the chief members of the *parlement*.

Chancellor Séguier was in turn assailed by a mob, finding refuge with his son-in-law, the duc de Luynes. Barricades went up in the streets, beginning the outbreak of the insurrection known as the Fronde. Presumably because he favored the government, Descartes quickly fled back to Holland. It was September 1, the same day as the death of his long acquaintance, Mersenne.[22] But Descartes was in such a hurry to leave that he would not even wait for the funeral.

A year later he would take ship for Stockholm. The regency had survived, to make peace with the Frondeurs in April 1649. The tomb raised for Descartes would prominently highlight the name of the queen regent of France; in 1666, the ceremonial translation of his bones back to Paris would—could it simply be a coincidence?—be performed under the watchful eyes of Anne's loyal servants.[23]

*

There is a bit of irony in someone who had been exiled later gaining a reputation as a representative of France. But at the same time representations of him avoided connecting him with the state itself, looking instead for something like the cosmopolitan spirit of French philosophy. Perhaps because Descartes's death occurred abroad, before he earned a triumphal return to Paris or great honors, images of him would celebrate not his loyalty but his self-aware, independent, critical "reason," something all people share. He had been knocked about, and died far from home, but fate had more than once kept him from being publicly identified with any particular party.

Yet when people read him, they often found Descartes's views rather acidic. He countered skepticism yet raised doubts about the provisional nature of many things we think we know; he established true knowledge on the basis of intuitive reason and yet advocated empirical studies of nature; he reassured those of faith that we have immortal souls but also showed that we are moved by passions so strong that few persons have any hope of controlling them, anchoring us in our bodies; he argued that we have volition and yet that the laws of nature are so immutable that not even God could change them. His youth had been full of materialist libertine associations, his young adulthood with the questions being raised by military engineers, Beeckman, and Galileo about nature as number and motion, his mature years mixed with friends Calvinist, Jansenist, and Oratorian, who all argued for an Augustinian providentialism, a charismatic faith and charity, and a deference to an embodied God born from Mary. He signed the baptismal record for his short-lived daughter in the chief Reformed church in Deventer and had to defend his faith in a letter to Mersenne since people back home were wondering if he had left the church (*de quelle religion il était*).[24] But later, the custodians of his reputation in France could make him into a conforming Catholic and even reconcile him with the militancy of Louis XIV's France; in yet another era he would metamorphose into a hero of revolutionaries who worshipped Nature.[25] Descartes's religious and ideological views were adaptable because he based them on a close knowledge of the three-dimensional world of human life rooted in a world of passion.[26] He and his readers sought universal knowledge based on unchanging truths, and yet they wished to grasp a world that was in constant motion, just like his own life: never stable nor fixed. But in the decades after his death, any such lines of argument would have been considered highly unorthodox in the increasingly regimented Catholicism of the reign of Louis XIV. Clerselier could not, then, quite work out how to present an acceptable version of his life; Baillet finally found a way to make him conformable, but only by dropping hints about his personal connections and aims, hints which have often been overlooked.

In the end, then, Descartes never simply became a courtier, nor simply a soldier, nor simply a philosopher. He was an active participant in contemporary events while also being keenly attentive to the phenomena of nature and human nature. He wished to understand the underlying commonalities among all people, which arose from a world in motion. What moved people was as real as anything, although the ways in which they expressed those movements were bound by custom. The underlying universal principles that govern us as well as all the rest could be grasped according to the three-dimensional physicomathematics with which he had become comfortable as a military engineer, founded in intuitive and instrumental truths. Descartes therefore stood among those who had reasons to support the view that we are made from the same matter as the rest of the universe, the corpuscularian motions of which could be understood without recourse to such fanciful abstractions as qualities or faculties. Sometimes he also wrote of an immaterial, rational soul, mingled with the extended world in the same way that gravity was mingled with matter, but he would not specify the properties of our volition nor any path toward personal salvation, placing his faith instead in Providence. He may have hoped for a millenarian moment, the return of a golden age, and he did what he could in person to aid those fighting on the front lines to reunite a deeply divided Europe. In the meantime, as he knew well, terrible suffering was burning through the lands. Only if one kept one's mind focused on clear and distinct ideas, grounded in the real world rather than in our imagination of it, would improvement for human life be possible. Morality began not by joining in grand designs or ideological positioning, but by doing one's best, with what was in front of you in light of the honest truth, and remaining loyal to the promptings of the true heart. Chivalry remained his code.

Yet the aggressive reactionaries of the day were using the shadows to mobilize hopes and fears built on doctrinal opinion and corporal discipline. No single person could stand up to the spies, propaganda, prisons, chains, poisons, and armies of absolutists in church and state. Descartes understood how to pummel well-built walls into rubble, a collective business that took planning, persistence, and matériel even more than élan. Knowledge of how to make such things happen circulated rapidly across regional divides, becoming a point of common knowledge about how physical matter responded to force with regularity.[27] If one just kept to the clearly and distinctly known real world, then, the means for cooperative problem solving were already to hand. The search for certainty demanded daily investment of time, resources, and attention, but the results would yield a clear-eyed view of our nature that could support material betterment and establish a more certain foundation for justice. The Descartes who had been brought up among *politiques*, libertines, and engineers was now an exile, searching for the means for making the real truths of the world known to all. Defending himself with his sword would not suffice. The mountain towered above, visible throughout the region. His friends urged him on and gave him aid. Others had failed to climb it, but he had experience, knowledge, and duty. Honor and affection demanded he try. He would find a way.

CHRONOLOGICAL TABLE

Young Descartes, 1596–1631

Year	Events in Descartes's Life	Events in France	Other Events
1596	March 30: Born at La Haye	King Henri IV, formerly Henri of Navarre, is reigning monarch	
	April 3: Baptized in the Catholic Church		
1597	May 13: Death of mother, Jeanne Brochard		
1598		Edict of Nantes grants privileges to Huguenots and ends France's civil wars	
1600	Father Joachim Descartes remarries, to Anne Morin, moves to Rennes	Henri IV marries Marie de Medici	
1601		Henri IV and Marie de Medici's son, Louis, born	
1603		King Henri IV lifts suspension of Jesuits in France	
1604		Discovery of spy network (Joachim Descartes probably involved)	
		Founding of Jesuit Collège Royale, at La Flèche	

Year	Events in Descartes's Life	Events in France	Other Events
1605		Protestants present grievances in Châtellerault	
1606	(or 1607?) Enters Collège Royale, La Flèche		
1607		Armand Jean du Plessis (later known as Richelieu) becomes Bishop of Luçon	
1609			Jülich-Cleves crisis (to 1610)
1610		May 13: Marie de Medici crowned queen May 14: Henri IV assassinated by a religious fanatic; eight-year-old Louis XIII becomes king, with queen mother as regent June 4: Heart of Henri IV received in La Flèche	
1612	November (?): Leaves school; living at father's house in Rennes (until March 1613?)		Matthias elected Holy Roman emperor
1613	Probably in Paris		
1614		October 2: Louis XIII's coming of age proclaimed	
	Late October: At meeting of Estates-General (through February 1615)		
1615		November: Louis XIII marries Anne of Austria	
1616	May 21: Signs as a godfather in Poitiers		
		September: Richelieu becomes minister of war and foreign affairs	
	November 9, 10: Takes law degree and license in Poitiers (then returns to Paris)		
1617		April 27: Palace coup of Louis XIII against Marie de Medici	

Year	Events in Descartes's Life	Events in France	Other Events
	Departs Paris		
			June: Ferdinand granted title of King of Bohemia
	October 22: Signs baptismal form in diocese of Nantes		
	December 3: Signs baptismal form in diocese of Nantes		
	Late December (?): In the Dutch Republic		
1618			February: Maurice becomes Prince of Orange
	March 31: Reaches the age of majority		
			May 23: Defenestration of Prague
			Summer: Maurice disarms Dutch city militias
			August 8: Order for arrest of Johan van Oldenbarnevelt
	November 10: Meets Isaac Beeckman in Breda		
			November 13: Synod of Dort (through April 23, 1619)
	December 31: Sends Treatise on Music to Beeckman		
1619	January 1: Begins "Parnassus" notebook		
		February 9: Execution of Giulio Cesare Vanini at Toulouse	
			March 20: Death of Emperor Matthias
	April 29: Intends to embark from Amsterdam for Copenhagen on the way to Hungary		

Year	Events in Descartes's Life	Events in France	Other Events
			May 13: Execution of Oldenbarnevelt
	July: Leaves Breda for Frankfurt (according to seventeenth-century biographer Adrien Baillet)		
		August 20: Treaty of Angoulême between Louis XIII and Marie de Medici	
			August 28: Ferdinand elected Holy Roman emperor
			September 9: Ferdinand crowned
			September 29: Frederick accepts crown of Bohemia
	Takes service with Maximillian of Bavaria		
	November 10–11: Has sequence of three dreams		
	Wintering in Neuburg (?)		
1620			May: Protestant Union musters at Ulm
	June 6: At Ulm, greets French delegation		
			July 3: Truce agreed in Germany
			August 25: Gabriel Bethlen, Prince of Transylvania, elected King of Hungary
	October: Traveling from Vienna to Prague (?)		
	November 8: Present at battle of White Mountain outside Prague		

Year	Events in Descartes's Life	Events in France	Other Events
1621			
	March: Takes service with count of Bucquoy		
		April: Huguenot conflict in France (through October 1622)	April: End of Dutch–Spanish truce
			April 15: Frederick and Elizabeth arrive at The Hague
			May 5: Bucquoy takes Bratislava
			July 10: Bucquoy dies at Neuhäsel / Nové Zámky
	July 27: Arrives in Pressburg, Bratislava		
	End of July: Heads north		
	On the Baltic coast, including Stettin		
	November: Begins return trip; incident onboard ship (?)		
	Winters in Dutch Republic		
1622			Bethlen settles with imperial forces
			February: Ambrogio Spinola captures Jülich after seige
	March: Receives inheritance from mother's estate		
	April 3: Signs legal document in Rennes	April 19: Richelieu made cardinal	
	May: In Poitou (?)	Spring: Royal forces in Nantes and Poitou	
1623	In Paris (?)		
		February: Treaty of Lyons over Valtellina signed	
	March 21: Writes letter to his brother, probably from Paris		

Year	Events in Descartes's Life	Events in France	Other Events
	April–July: In Poitou, or with court?		
	June–July Sales of estates in Poitou		
	August: Probably visits Paris and then heads to Alps	August: Parlement of Paris sentences Théophile de Viau in absentia	August 6: Urban VIII elected pope
			October 27: Galileo's *Assayer* presented to Urban VIII
	Winter traveling in Alpine region		
1624		April: Richelieu made royal counsellor	
	May: In Venice		
	Spring: To Loreto (?)		
	Summer: In Valtellina (?)		
		August: Richelieu becomes chief adviser to Louis XIII	
	Late November: In Rome	Late November: French army attacks in Valtellina	
1625		Huguenot conflict in France (through early 1626)	
		February: France and Savoy attack Genoa (through March 1626)	
	March: Begins return to Paris in company of Cardinal Francesco Barberini; then with French forces	March–September: Cardinal Barberini in Paris	
		April 22: French take Gavi	
		May–June: Henrietta Maria's marriage by proxy to King Charles I, journey to England	
	June 24: Writes letter to father from Poitou, exploring office in Châtellerault; then returns to Paris		
		September: Théophile de Viau acquitted	

Year	Events in Descartes's Life	Events in France	Other Events
1626		May 3: *Maréchal* (marshal) d'Ornano arrested in plot against Richelieu	
	Summer: In Poitou and Rennes		
		July 8: Arrest of comte de Chalais	
	July 16: Writes letter to his brother from Paris		
		August 6: Wedding of Gaston d'Orléans and Marie de Bourbon, duchesse de Montpensier	
		August 19: Execution of Chalais	
	Return to Paris; meetings at Nicolas Le Vasseur's (?)		
1627			July: Duke of Buckingham's expedition to Isle de Ré
		August: Siege of La Rochelle begins	
	November 8: "Le cap^{ne} Descart" involved in action at Isle de Ré		
1628	January 22: Appears as godfather to nephew, in Brittany		
	Defense of friend Guez de Balzac (?)		
		March 30: Balzac writes letter to Descartes, when Descartes is departing Brittany	
		May 15: English relief fleet fails to relieve La Rochelle	
			August: Assassination of Duke of Buckingham
	Summer: Attending siege of La Rochelle		
		September: Another English relief fleet fails	

Year	Events in Descartes's Life	Events in France	Other Events
	October 8: Visits Beeckman in Dordrecht		
	October 28: Enters La Rochelle with French royal occupation forces; siege ends		
	November (?): Attends meeting at papal nuncio's		
1629		February: Louis XIII moves to relieve Casale	
	March: Meets with Beeckman in Dordrecht		
	April: Matriculates at university of Franeker		
		May: France and England settle conflict	May: Frederick Henry besieges s'Hertogenbosch (through September 17)
		June 28: Peace of Alais ends Huguenot conflict	
		October: Bérulle dies	
1630	Declines to travel to Constantinople, probably makes brief visit to England		
	June 27: Matriculates at university of Leiden		
		November 10: "Day of the Dupes" ends with Richelieu retaining power	
1631		March: Balzac publishes *Le Prince* in praise of Richelieu	
	April 15: Writes letter to Balzac, restoring communication, but does not return to France		
	May (?): Makes trip to Denmark with Étienne Villebressieu, returns to Amsterdam		

EARLY CORRESPONDENCE AND PUBLICATIONS

Early Surviving Correspondence

From project on Circulation of Knowledge and Learned Practices in the 17th Century Dutch Republic, Huygens Institute (http://ckcc.huygens.knaw.nl). Also available through Early Modern Letters Online, http://emlo.bodleian.ox.ac.uk/forms/advanced?people=descartes

1619	24 Jan	Descartes (Breda) to Beeckman (Middelburg)
	9 Feb	Descartes (Breda) to Beeckman (Middelburg)
	26 Mar	Descartes (Breda) to Beeckman (Middelburg)
	20 Apr	Descartes (Breda) to Beeckman (Middelburg)
	23 Apr	Descartes (Breda) to Beeckman (Middelburg)
	29 Apr	Descartes (Amsterdam) to Beeckman (Middelburg)
	6 May	Beeckman (Middelburg) to Descartes (Copenhagen)
1622	22 May	Descartes (?) to father, Joachim (?) [missing; known to Baillet]
1623	21 Mar	Descartes (Paris) to brother, Pierre [missing; known to Baillet]
1625	24 Jun	Descartes (Poitiers) to father, Joachim (?) [missing; known to Baillet]
	?	Mydorge (Paris) to Descartes (Paris)
1626	Feb	Descartes (Paris) to Mersenne (Paris)
	16 Jul	Descartes (Paris) to brother, Pierre (?) [missing; known to Baillet]
1628	30 Mar	Balzac (Paris) to Descartes (Brittany?)
	?	Descartes (?) to unknown male (?)

1629	18 Jun	Descartes (Franeker) to Ferrier (Paris)
	18 Jul	Descartes (Franeker) to Gibieuf (Paris)
	Aug	Descartes (Franeker) to Mersenne (Paris)
	Sep	Descartes (Franeker) to unknown male (?)
	8 Oct	Descartes (Amsterdam) to Mersenne (Paris)
	8 Oct	Descartes (Amsterdam) to Ferrier (Paris)
	26 Oct	Ferrier (Paris) to Descartes (Amsterdam)
	13 Nov	Descartes (Amsterdam) to Ferrier (Paris)
	13 Nov	Descartes (Amsterdam) to Mersenne (Paris)
	20 Nov	Descartes (Amsterdam) to Mersenne (Paris)
	18 Dec	Descartes (Amsterdam) to Mersenne (Paris)
. . .		
1631	15 Apr	Descartes (Amsterdam) to Balzac (Paris)
	25 Apr	Balzac (Paris) to Descartes (Amsterdam)
	5 May	Descartes (Amsterdam) to Balzac (Paris)
. . .		

Chief Publications during Descartes's Lifetime (1596–1650) and Shortly Thereafter

1637: *Discours de la method . . . la Dioptrique, les Meteores, et la Geometrie*

1641: *Meditations de prima philosophia*; 2nd ed. 1642; French trans. 1647

1644: *Principia philosophiae*; French ed., *Les principes de la Philosophie*, 1647

1649: *Les passions de l'âme*

1662: *De Homine*; French ed., *L'Homme*, 1664; with *La description du corps humaine*

1664: *Le Monde, ou Traité de la Lumiere*

ACKNOWLEDGMENTS

All authors are indebted to those who have helped (and to those who by hindering have sharpened the point). This project on Descartes before he became a published author took form only recently, although understanding Descartes as a Dutch philosopher had been an aspect of an earlier book. In retrospect, questions had been building in my mind for some decades, no doubt from conversations with David Lux, Hilton Root, Carol Lansing, and other fellow students during my graduate student years and the mentorship of Toon Kerkhoff and especially Harm Beukers during the first period when I was in The Netherlands learning about early modern Dutch medicine and science. More recently, I have been inspired by the work and rare conversation of Chandra Mukerji, who has opened my eyes more than anyone to the importance of military engineering and impersonal governance in Descartes's period, which provided a foundation for thinking about his work afresh.

My deepest debt, however, is to the students I have had, undergraduates as well as graduate students, whose questions and answers helped me confront some of the usual historical accounts I was passing on about commonly named persons and ideas. I have come to appreciate the importance of teaching-led research, which requires course leaders continually to reexamine what they know, what they need to explore further, and how it might make sense in terms of other parts of knowledge. These days we often hear about the importance of justifying our craft through public engagement, which often takes the form of telegraphing exciting conclusions; but for instructors the first and daily engagement with students stimulates investigations over many years with attention to particular questions and problems as a part of sustained conversations. Attentive study brings light. It is a real privilege to have been able to work among alert colleagues and students over the past three and a half decades, most recently at Brown University, and to have been supported in recent years by one of its former students, John F. Nickoll, who has generously given to sustain the enterprise.

Recently, three especially sharp and energetic students at Brown University who took a chance to enroll in a seminar on Descartes and His Age helped me understand what they wanted and needed to know: Anna Martin, Gina Milano, and Rebecca Millstein. Their enthusiasm for the subject encouraged me to keep looking. A chance to speak with Paul Knoll and other residents of the Mirabella in Portland, Oregon, was also very encouraging. I am especially grateful to the early support offered by family, friends, and colleagues when I tried out on them the idea of writing a book on Descartes before he became a philosopher.

Other important aid came from chances to explore the legacy of Descartes and his work in academic conferences or workshops. One of my first outbursts about my unhappiness with the usual account was expressed about a decade ago at a large public forum hosted by the astute but patient scholars of the Descartes Centre in Utrecht, not far from where some of the first confrontations about Descartes's views were shouted almost four centuries ago. A request a few years ago from Larry Nolan to write a short summary on "Medicine" for *The Cambridge Descartes Lexicon* prompted me to work through some of Descartes's writings more systemically than I had done before. I am thankful to the program committees of three academic associations who agreed to let me speak on aspects of Descartes at their annual meetings: the History of Science Society, the American Association for the History of Medicine, and the Renaissance Society of America. I am also grateful for invitations to speak when early parts of the book were coming together, particularly the Oslo Medical History Institute, the Institute for the History of Medicine at Johns Hopkins University, the Institute for Research in the Humanities at the University of Wisconsin–Madison, the Consortium for the Study of the Premodern World at the University of Minnesota, the Early Sciences Working Group and the Early Modern History Workshop of Harvard University, and the Department of History of Reed College.

Opportunities for periods of time devoted to reading and research are also crucial for any author. My special thanks, for a four-month fellowship, are due to the Koninklijke Bibliotheek in The Hague and the Netherlands Institute for Advanced Study in Wassenaar, and to the staff and fellows who made up such a lively community, at the end of which Faye and I were able to rent a car and travel to many of the northern European places associated with Descartes.

This being a project almost entirely based on printed materials, I extend my gratitude to the librarians and staff of the Rockefeller and Hay libraries at Brown University, especially those who efficiently handled all the Interlibrary Loan requests. Academic colleagues and acquaintances were also generous not only with supportive goodwill but also with suggestions for reading that helped in countless ways. I am especially grateful to the knowledgeable assistance offered by Maria Pia Donato, Robert Schneider, Daniel Garber, Joan Richards, Jim Bono, and David Sacks, as well as to conversations with Theo Verbeek, Catherine Wilson, Brooke Holmes, Stephen Gaukroger, Malcolm Smuts, Ofer Gal, Tonba Ghadessi, Evelyn Lincoln, Caroline Castiglione, and Tara Nummedal. Lyse Mesmer has taken some prompts and turned them into wonderful working maps. Thanks to Lori Meek Schuldt for expert copyediting, and to Eileen Quam for an intelligent index. Karen Merikangas Darling, as editor of works in the history and philosophy of the sciences, encouraged this project and helped shape its framing; she also obtained helpful opinion from referees. The three reviews she obtained were enormously helpful. I remain indebted to the systems of review that support scholarly publication and which rely mainly on the shared generosity of writerly communities. Errors of fact and interpretation remain my own responsibility.

One person has patiently taken onboard all the absentmindedness and ups and downs that come with long-term projects and even helped me chase the ghost of Descartes on the ground, sustaining her constant support and love throughout, without whom this could not have been completed: Faye.

FOR FURTHER READING

The chief early biographies of Descartes and more recent studies are discussed in part 1 of this work. Additional details are to be found in the notes. But while many English-language readers with further questions will these days understandably look to Wikipedia for information, anyone looking for richer, more reliable narratives of this period might like to turn to a few older or encompassing introductions to people and events:

Batiffol, Louis. *The Duchesse de Chevreuse: A Life of Intrigue and Adventure in the Days of Louis XIII*. Translated by Carrie Chapman Catt. New York: Dodd, Mead, 1914. First published in 1913.

Burckhardt, Carl Jacob. *Richelieu and His Age*. 3 vols. Translated by Edwin Muir, Willa Muir, and Bernard Hoy. London: Allen and Unwin, 1967. First published in 1891.

Pardoe, Julia. *The Life of Marie de Medicis, Queen of France, Consort of Henry IV, and Regent of the Kingdom under Louis XIII*. 6 vols. New York: James Pott, 1905. First published in 1852.

Parker, Geoffrey. *Global Crisis: War, Climate Change and Catastrophe in the Seventeenth Century*. New Haven, CT: Yale University Press, 2013.

Tapié, Victor-L. *France in the Age of Louis XIII and Richelieu*. Translated by D. McN. Lockie. London: Macmillan, 1974. First published in 1967.

Wilson, Peter H. *The Thirty Years War*. Cambridge, MA: Belknap Press of Harvard University Press, 2009.

NOTES

Part One

1. Members of Descartes's family bore titles such as *ecuyer* (squire), and they clearly thought they deserved that status: it may have been the lowest of noble titles, but a noble title nonetheless; they later gained the right to use the term *chevalier* (knight). The family's coat of arms and title of *chevalier* are clearly evident in the engraving of Gerard Edelinck, printed as the frontispiece for Adrien Baillet, *La Vie De Monsieur Des-Cartes*, 2 vols. (Paris: Chez D. Horthemels, 1691).

2. On his learning to ride the "great horse," see Baillet, *Des-Cartes*, 1:35; for habits of dress and love of jokes, see Charles Adam, ed., *Vie et Oeuvres de Descartes: Étude historique; Supplément a l'édition de Descartes*, vol. 12 of *Oeuvres De Descartes*, ed. Charles Adam and Paul Tannery (Paris: L. Cerf, 1910), 74; for his love of ancient mythology, not for its lessons but for its rousing stories, see Baillet, *Des-Cartes*, 1:19–20; on his gambling, see Baillet, *Des-Cartes*, 1:36. For the "fundamentally social" aims of Descartes's philosophy, see Peter Dear, "A Mechanical Microcosm: Bodily Passions, Good Manners, and Cartesian Mechanism," in *Science Incarnate: Historical Embodiments of Natural Knowledge*, ed. Christopher Lawrence and Steven Shapin (Chicago: Chicago University Press, 1998), 51–82.

3. Poisson writes that the papers were now in the hands of M. Clerselier (whom we will encounter subsequently): Nicholas-Joseph Poisson, *Commentaire ou remarques sur la methode de René Descartes* (1670; New York: Garland, 1987), 20–21. Also at Charles Adam and Paul Tannery, eds., *Oeuvres de Descartes*, 12 vols. (Paris: L. Cerf, 1897–1910), 10:255–56.

4. Pierre Borel, *The Life of the Most Famous Philosopher Renatus Descartes* (London: E. Okes et al., 1670), 7.

5. Stephen Snelders, *Vrijbuiters van de heelkunde: Op zoek naar medische kennis in de tropen 1600–1800* (Amsterdam/Antwerp: Atlas, 2012).

6. "Verae philosophiae, quam vocat operam navantium": from Beeckman's 1628 "Historia" of his relationship with Descartes, in C. Adam and Tannery, *Oeuvres de Descartes*, 10:332; the second metaphor is from the *Discours*; see John Cottingham, Robert Stoothoff, and Dugald Murdoch, eds., *The Philosophical Writings of Descartes*, 3 vols. (Cambridge: Cambridge University Press, 1985–1991), 1:145.

7. *Amadis of Gaul, Books I and II*, ed. Garci Rodríguez de Montalvo, trans. Edwin B. Place and Herbert C. Behm (Lexington: University Press of Kentucky, 2003), 11. On Descartes's acquaintance with *Amadis* and other romances, see C. Adam, *Vie et Oeuvres*, 73.

8. C. Adam and Tannery, *Oeuvres de Descartes*, 10:538. On a lost treatise on fencing written by Descartes, see Baillet, *Des-Cartes*, 1:35.

9. Borel, *Renatus Descartes*, 6–7.

10. Jeroen van de Ven, "Quelques données nouvelles sur Helena Jans," *Bulletin Cartésien* 31 (2003): 10–12.

11. Adam (*Vie et Oeuvres*, 236) thinks this must be the Duchesse d'Aiguillon; but it is an inference of Adam's from a reference in a letter of Descartes to Mersenne of 25 May 1637, thanking him for getting the permission to publish the *Discours*, being particularly indebted "à cette Dame qui vous a écrit, de ce qu'il luy plaist juger de moy si favorablement" ("To a certain lady who wrote to you, of which she was pleased to judge of me so favorably") (C. Adam and Tannery, *Oeuvres de Descartes*, 1:376). It might be another noble woman.

12. Erica Harth, *Cartesian Women: Versions and Subversions of Rational Discourse in the Old Regime* (Ithaca, NY: Cornell University Press, 1992).

13. Louis Batiffol, *The Duchesse de Chevreuse: A Life of Intrigue and Adventure in the Days of Louis XIII* (New York: Dodd, Mead, 1914). The text in the upper left of figure 2 states, "Ce cerf a esté laissé courre et pris au mont temery par charles duc de lorraine et de barle 15 julllet 1627" ("This deer has been hunted and taken on leash at Mont Temery by Charles duc de Lorraine and Barle, 15 July 1627"). Charles of Lorraine and the duchesse (then married to Charles's brother) were lovers at the time, so the portrait is clearly metaphorical. They were also plotting against Richelieu, and the timing coincides with Buckingham's landing on the Isle de Ré, leading to the siege of La Rochelle. The reference to Mont Temery is likely to be to a rise southeast of Cherbourg.

14. Geneviève Rodis-Lewis, *Descartes: His Life and Thought*, trans. Jane Marie Todd (Ithaca, NY: Cornell University Press, 1998), 25, wrestles with this but concludes that it was not exceptional.

15. Baillet, *Des-Cartes*, 1:168.

16. Maxime Leroy, *Descartes: Le Philosophe au Masque* (Paris: Rieder, 1929), 25, 108; René Pintard, *Le Libertinage Érudit dans la Première Moitié du XVIIe siècle*, 2 vols. (Paris: Boivin, 1943), 203–4.

17. Gustave Cohen, *Écrivains Français en Hollande dans le Première Moitié du XVIIe siècle* (Paris: Édouard Champion, 1920), 636.

18. Theo Verbeek, ed., *La Querelle d'Utrecht: René Descartes et Martin Schoock* (Paris: Les impressions nouvelles, 1988), 33; the treatise in question was the *Admiranda Methodus* of Martinus Schoock, translated in full into French at 157–320.

19. John Boswell, *Christianity, Social Tolerance, and Homosexuality: Gay People in Western Europe from the Beginning of the Christian Era to the Fourteenth Century* (Chicago and London: University of Chicago Press, 1980), 25; Nicholas Hammond, *Gossip, Sexuality and Scandal in France* (Oxford: Peter Lang, 2011); for the earlier example of Florence, Michael Rocke, *Forbidden Friendships: Homosexuality and Male Culture in Renaissance Florence* (Oxford: Oxford University Press, 1996); for a sensitive account of mainly English examples, see Alan Bray, *The Friend* (Chicago: University of Chicago Press, 2003).

20. From the *Passions* (1649), in Cottingham, Studoff, and Murdoch, *Philosophical Writings*, 1:357.

21. Mitchell Greenberg, *Subjectivity and Subjugation in Seventeenth-Century Drama and Prose: The Family Romance of French Classicism* (Cambridge: Cambridge University Press, 1992), chap. 1: "L'Astrée and Androgyny," 24–47; also see the introduction to Honoré d' Urfé, *Astrea (Part One)*, ed. and trans. Steven Rendall (Binghamton, N.Y.: Medieval and Renaissance Texts and Studies, 1995); and Leah DeVun, "The Jesus Hermaphrodite: Science and Sex Difference in Premodern Europe," *Journal of the History of Ideas* 69 (2008): 193–218.

22. Walther P. Fischer, *The Literary Relations Between La Fontaine and the "Astrée" of Honoré D' Urfé* (Philadelphia: Publications of the University of Pennsylvania Series in Romanic Languages and Literatures, 1913), 6.

23. The usual account was told since the first biography, by Borel (*Renatus Descartes*, 25); for a revised view of his meetings with the queen, see Susanna Åkerman, *Queen Christina of Sweden and Her Circle: The Transformation of a Seventeenth-Century Philosophical Libertine* (Leiden: E. J. Brill, 1991), 49; on the rumors of poisoning, see Åkerman, *Queen Christina*, 51; a recent argument in favor of the hypothesis, which I have not seen, is Theodor Ebert, *Der rätselhafte tod des René Descartes* (2009).

24. Thomas M. Lennon, ed., *Against Cartesian Philosophy* (Amherst, NY: Humanity Books, 2003), 31–32. The rumor may have been abroad from the time of his death, since Christiaan Huygens told Baillet that the *Gazette d'Anvers* [Antwerp] reported that a madman had died in Sweden so that Descartes could live as long as he wished: G. Cohen, *Écrivains Français en Hollande*, 404–5, quoting C. Adam and Tannery, *Oeuvres de Descartes*, 5:630.

25. Rodis-Lewis, *Descartes: Life and Thought*, 4.

26. Jon R. Snyder, *Dissimulation and the Culture of Secrecy in Early Modern Europe* (Berkeley: University of California Press, 2009); Françoise Viatte, Domi-

nique Cordellier, and Violaine Jeammet, eds., *Masques Mascarades Mascarons* (Paris: Musée du Louvre, 2014).

27. Anthony Studler van Zurck (who will be discussed subsequently), served as an early drop box for Mersenne.

28. See Edelinck's engraving, figure 1; he quotes it as his motto in a letter of 1634 to Mersenne (Cottingham, Stoothoff, and Murdoch, *Philosophical Writings*, 3:43).

29. It also appeared inscribed on one of the "indecent" or "bawdye" paintings of Johannes Torrentius, from the 1620s: A. Bredius, *Johannes Torrentius Schilder, 1589–1644* (The Hague: Martinus Nijhoff, 1909), 9.

30. John W. Montgomery, *Cross and Crucible: Johann Valentin Andreae (1586–1654): Phoenix of the Theologians*, 2 vols. (The Hague: M. Nijhoff, 1973).

31. For a summary of the possibility of Descartes' Rosicrucian associations, see William R. Shea, "Descartes and the Rosicrucian Enlightenment," in *Metaphysics and Philosophy of Science in the Seventeenth and Eighteenth Centuries: Essays in Honour of Gerd Buchdahl*, ed. R. S. Woolhouse (Dordrecht: Kluwer, 1988), 73–99.

32. Gabriel Daniel, *A Voyage to the World of Cartesius. Written Originally in French*, trans. T. Taylor (London: Thomas Bennet, 1692), 14–21; called belladonna.

33. Cottingham, Stoothoff, and Murdoch, *Philosophical Writings*, 1:111. On his self-conscious crafting of his work, knowing that not everyone would welcome his conclusions, see Amy Mullin, "If Truth Were Like Money: Descartes and His Readers," *History of Philosophy Quarterly* 19 (2002): 149–69.

34. Quoted in Cottingham, Stoothoff, and Murdoch, *Philosophical Writings*, 1:111, although I have substituted "person" for "man."

35. Ibid., 1:112.

36. Ibid., 3:52, letter of February 27, 1637.

37. Baillet, *Des-Cartes*, xxviii–xxx.

38. See the introduction to Theo Verbeek, Erik-Jan Bos, and Jeroen van de Ven, eds., *The Correspondence of René Descartes 1643* (Utrecht: Zeno, 2003).

39. On Hogelande's alchemical work, see Bernard Joly, *Descartes et la Chimie* (Paris: J. Vrin, 2011), 67–69.

40. Quoted from René Descartes to Cornelis van Hogelande, 30 August 1649, in Baillet, *Des-Cartes*, xxviii–xxix.

41. On De Raey's opinion about the letters: "trés petit nombre & de peu d'importance: & que M. Descartes avoit emporté les principaux en Suéde," ibid., xxviii. For his comment on the French, see ibid., xxx.

42. C. Adam and Tannery, *Oeuvres de Descartes*, 5:406–9; G. Cohen, *Écrivains Français en Hollande*, 677; the amount was the huge sum of 9,000 livres of Dutch money, the equivalent of 10,500 livres of French money. Van Zurck also possessed a copy of the manuscript later published by Schouten as *De Homine*: see G. A. Lindeboom, *Florentius Schuyl (1619–1669) en zijn Betekenis voor het Cartensianisme in de Geneeskunde* (The Hague: Martinus Nijhoff, 1974), 70.

43. Verbeek, Bos, and van de Ven, *Correspondence 1643*, xi–xv.

44. "M. [Johannes] de Raey pouroit bien avoir été cet ami discret à qui M. de Hooghelande auroit fait lire des lettres avant que de les brûler: & si elles n'ont pas été brûlées, il n'y a peut-être eu que la crainte de les rendre utiles au Public qui luy en a fait faire un mystére à M. Van Limborch." Baillet, *Des-Cartes*, 1:xxviii–xxix.

45. C. Adam and Tannery, *Oeuvres de Descartes*, 10:1–14.

46. Baillet, *Des-Cartes*, 1:xvi–xxi; also see John R. Cole, *The Olympian Dreams and Youthful Rebellion of René Descartes* (Urbana: University of Illinois Press, 1992), 21–30.

47. Baillet, *Des-Cartes*, 1:xxxi.

48. Leibniz's *Cogitationes privatae* of 1676, in which he took extensive notes on Descartes's now missing "Little Notebook" (or *Olympica*), as paraphrased and translated in Baillet, *Des-Cartes*, 1:50–51, 81–86: all versions are in Adam and Tannery, *Oeuvres de Descartes*, 10:5–12 (Chanut), 211–19 (Leibniz), and 179–88 (Baillet).

49. Baillet, *Des-Cartes*, 1:xiv–xv; also see Adrien Baillet, *Jugemens des Savans sur les Principaux Ouvrages des Auteurs*, 7 vols. (Paris: Charles Moette et al., 1722), 1:19, where he says that Lipstorp came to Paris for several years and knew Baillet well, giving him everything he published and much else, including a work on how to decode the scriptures for prognostication, which Baillet translated into French in 1688.

50. Baillet, *Des-Cartes* t, xv–xvi; Borel also published *Discours nouveau prouvant la pluralité des mondes*, arguing that the moon and stars were inhabited; this is an Epicurean many-worlds theory, further suggesting why he was interested in Descartes's philosophy.

51. Chandra Mukerji, *Impossible Engineering: Technology and Territoriality on the Canal du Midi* (Princeton, NJ: Princeton University Press, 2009), 43.

52. See Baillet, *Des-Cartes*, xii–xiii.

53. Charles Adam thought that Baillet had access to memoirs of Descartes left by Clerselier, Chanut, Mydorge, Hardy, La Vasseur, De la Barre, and Auzout (Adam, *Vie et Oeuvres*, v).

54. Baillet, *Jugemens des Savans*, the "Abregé de la vie de Mr. Baillet" by B. La Monnoye is 1:3–28.

55. Baillet, *Des-Cartes*; Adrien Baillet, *The Life of Monsieur Des Cartes, Containing the History of His Philosophy and Works*, trans. S. R. (London: Printed for R. Simpson, at the Harp in St. Paul's Church-yard, 1693); also see the appreciation of Baillet in Cole, *Olympian Dreams*, 41–48.

56. Verbeek, Bos, and van de Ven, *Correspondence 1643*, xxiv–xxv; also Adam, *Vie et Oeuvres*, iii–v.

57. Rodis-Lewis, *Descartes: Life and Thought*, xvi; Åkerman, *Queen Christina*, 35.

58. Alain Tallon, *La Compagnie du Saint-Sacrement (1629–1667): Spiritualité et Société* (Paris: Cerf, 1990), 108.

59. Ibid., 119; Benoist Pierre, *La Monarchie ecclésiale: Le clergé de cour en France à l'époque moderne* (Seyssel: Champ Vallon, 2013), 350.

60. Ruth Kleinman, *Anne of Austria: Queen of France* (Columbus: Ohio State University Press, 1985), 253.

61. René de Voyer d'Argenson, *Annales de la Compagnie du Saint-Sacrement*, ed. H. Beauchet-Filleau (Marseille: Saint-Léon, 1900); Raoul Allier, *La Cabale de Dévots: 1627–1666* (Paris: Armand Colin, 1902).

62. Baillet, *Des-Cartes*, dedication.

63. Rodis-Lewis, *Descartes: Life and Thought*, xiii, notes that because of this criticism of sources, Baillet was suspected by the Jesuits of being a Jansenist.

64. On the *Jugements des Savans*, see April Shelford, *Transforming the Republic of Letters: Pierre-Daniel Huet and European Intellectual Life, 1650–1720* (Rochester, NY: University of Rochester Press, 2007), 170; for his meditations on authorial disguise, see Nick Wilding, *Galileo's Idol: Gianfrancesco Sagredo and the Politics of Knowledge* (Chicago: University of Chicago Press, 2014), 129; for evidence of Baillet as a copyist, Cole, *Olympian Dreams*, 41–48.

65. The Mersenne correspondence was published in seventeen volumes between 1932 and 1987.

66. The recent edition of the Descartes correspondence is being produced through the Circulation of Knowledge project in the Netherlands under the editorial supervision of Eric-Jan Bos and can be accessed through the Early Modern Letters Online project hosted by Oxford University: http://emlo-portal.bodleian.ox.ac.uk/collections/?catalogue=rene-descartes#partners.

67. For instance, the edition reprinted by Vrin in 1964.

68. The quotation comes from a philosopher who came to be quite unhappy with modernism: Stephen Toulmin, *Cosmopolis: The Hidden Agenda of Modernity* (Chicago: University of Chicago Press, 1992), 108.

69. Roger Ariew et al., eds., *Historical Dictionary of Descartes and Cartesian Philosophy* (Lanham, MD: Scarecrow Press, 2003).

70. Ibid., all quotations from 3–4.

71. Geneviève Rodis-Lewis, *Descartes: Biographie* (Paris: Calmann-Lévy, 1995); Rodis-Lewis, *Descartes: Life and Thought*.

72. Stephen Gaukroger, *Descartes: An Intellectual Biography* (Oxford: Clarendon Press, 1996); see also Desmond Clarke, *Descartes: A Biography* (Cambridge: Cambridge University Press, 2006), Daniel Garber, *Descartes Embodied: Reading Cartesian Philosophy through Cartesian Science* (Cambridge: Cambridge University Press, 2001), and especially John Andrew Schuster, *Descartes-Agonistes: Physico-Mathematics, Method and Corpuscular-Mechanism 1618–33* (Dordrecht: Springer, 2013). For example, see Gaukroger's *Descartes*, 6; he also—correctly, in my opinion—chastises Alexandre Koyré for trying to turn the history of science into the history of epistemology (p. 14). For a similar attack on the history of science as metaphysics, see Gary Hatfield, "Metaphysics and the New Science," in *Reappraisals of the Scientific Revolution*, ed. David C. Lindberg and Robert S. Westman (Cambridge: Cambridge University Press, 1990).

73. Toulmin, *Cosmopolis*, 56–62. Adam discusses the ceremony and the poem (*Vie et Oeuvres*, 28–31), although he does not attribute it to Descartes. Rodis-Lewis (*Descartes: Life and Thought*, 14) considers Toulmin's attribution "plausible."

74. A. C. Grayling, *Descartes: The Life of René Descartes and Its Place in His Times* (London: Pocket Books, 2006), 9, 11.

75. Ibid., 46–47.

76. J. P. D. Cooper, *The Queen's Agent: Francis Walsingham at the Court of Elizabeth I* (London: Faber, 2012); more generally, James Westfall Thompson and Saul Kussiel Padover, *Secret Diplomacy: Espionage and Cryptography, 1500–1815* (New York: F. Ungar, 1963).

77. Jacob Soll, *The Information Master: Jean-Baptiste Colbert's Secret State Intelligence System* (Ann Arbor: University of Michigan Press, 2009); Filippo De Vivo, *Information and Communication in Venice: Rethinking Early Modern Politics* (Oxford: Oxford University Press, 2007).

78. H. R. Trevor-Roper, *Europe's Physician: The Various Life of Sir Theodore De Mayerne* (New Haven, CT: Yale University Press, 2006); John Bossy, *Giordano Bruno and the Embassy Affair* (New Haven, CT: Yale University Press, 1994); Pamela Smith, *The Business of Alchemy: Science and Culture in the Holy Roman Empire* (Princeton, NJ: Princeton University Press, 1994); Alice Stroup, "Nicolas Hartsoeker, savant Hollandias associé de l'académie et espion de Louis XIV," in *De la diffusion des sciences à l'espionnage industriel XVe–XXe siècle*, ed. André Guillerme (1999); "Voltaire, agent secret du roi," Histoire Pour Tous, February 6, 2013, http://www.histoire-pour-tous.fr/histoire-de-france/4464-voltaire-agent-secret-du-roi.html; Steven Nadler, *Spinoza: A Life* (Cambridge: Cambridge University Press, 1999), 314–19.

79. Marika Keblusek and Badeloch Noldus, eds., *Double Agents: Cultural and Political Brokerage in Early Modern Europe* (Boston: Brill, 2011).

80. Richard Watson, *Cogito, Ergo Sum: The Life of René Descartes* (Boston: David R. Godine, 2002).

81. G. Cohen, *Écrivains Français en Hollande*, 645; Frederick Henry had died on March 14, 1647.

82. Herbert H. Rowen, *John de Witt, Grand Pensionary of Holland, 1625–1672* (Princeton, NJ: Princeton University Press, 1978), 45.

83. While working on a previous book, which included a chapter on Descartes (Harold J. Cook, *Matters of Exchange: Commerce, Medicine and Science in the Dutch Golden Age* [New Haven, CT: Yale University Press, 2007]), I also came to think that he might have been a spy, and around 2002 I spoke to friends in London about it; I later gave papers on the possibility, in Utrecht and the Renaissance Society of America. These were early steps toward this work.

84. Grayling, *Descartes: Life and Its Place*, 11.

85. Ibid.

86. Watson, *Cogito*, 22–23, writing in the tradition of Leroy, *Descartes: Philosophe au Masque*.

87. Watson, *Cogito*, 153.

88. Ibid., 147, 150.

89. Quoted by Rodis-Lewis, *Descartes: Life and Thought*, xv, from Adam, *Vie et Oeuvres*, 305.

90. G. Cohen, *Écrivains Français en Hollande*; the section on Descartes makes up the second half of the book, 357–689.

91. René Descartes to Pierre Chanut, 1 February 1647, in Adam, *Vie et Oeuvres*, 78.

92. Leroy, *Descartes: Philosophe au Masque*; Dimitri Davidenko, *Descartes le Scandalleux* (Paris: Robert Laffont, 1988). See also Anne Staquet, *Descartes et le Libertinage* (Paris: Hermann, 2009).

93. Pintard, *Libertinage Érudit*, 203–4.

94. See, for instance, Françoise Charles-Daubert, *Les libertins érudits en France au XVIIe siècle* (Paris: Presses Universitaires de France, 1998); Anthony McKenna and Pierre-François Moreau, eds., *Libertinage et philosophie au XVIIe siècle: La résurgence des philosophies antiques* (Saint-Étienne, France: L'Université de Saint-Étienne, 2003); Didier Foucault, *Un Philosophe libertin dans l'Europe Baroque: Giulio Cesare Vanini (1585–1619)* (Paris: Honoré Champion, 2003); Marcella Leopizzi, *Les sources documentaires du courant libertine Français: Giulio Cesare Vanini* (Fasano, Italy: Schena Editore; Paris: Presses de l'Université de Paris-Sorbonne, 2004); Jean-Pierre Cavaillé, *Dis/simulations: Jules-César Vanini, François La Mothe Le Vayer, Gabriel Naudé, Louis Machon et Torquato Accetto: Religion, morale et politique au XVIIe siècle* (Paris: Honoré Champion, 2008).

95. Charles-Daubert, *Les libertins érudits*, 97.

96. Catherine Wilson, "Descartes and the Corporeal Mind: Some Implications of the Regius Affair," in *Descartes' Natural Philosophy*, ed. Stephen Gaukroper, John Schuster, and John Sutton (London: Routledge, 2000), and Catherine Wilson, *Epicureanism at the Origins of Modernity* (Oxford: Clarendon Press, 2008); Åkerman, *Queen Christina*; Staquet, *Descartes et le Libertinage*; Alexandra Torero-Ibad, *Libertinage, Science et Philosophie dans le matérialism de Cyrano de Bergerac* (Paris: Honoré Champion, 2009).

97. Russell Shorto, *Descartes' Bones: A Skeletal History of the Conflict between Faith and Reason* (New York: Doubleday, 2008); Steven Nadler, *The Philosopher, the Priest, and the Painter: A Portrait of Descartes* (Princeton, NJ: Princeton University Press, 2013); Tad M. Schmaltz, *Radical Cartesianism: The French Reception of Descartes* (Cambridge: Cambridge University Press, 2002).

98. A good place to see this at work is in Natalie Zemon Davis, *Trickster Travels: A Sixteenth-Century Muslim between Worlds* (New York: Hill and Wang, 2006). My thanks to Caroline Castiglione for the conversation that brought up the phrase "tangential evidence."

Part Two

1. D'Argenson, "Note sur la famille Descartes et l'origine de son name," *Mémoires de la Société archéologique de Touraine* 4 (1847): 87–94.

2. C. Adam, *Vie et Oeuvres*, 590–91: on the left face of his tomb, "Renatus Des-Cartes, Perronij Dominus, &c. / Ex Antiquâ & Nobili inter Pictones & Armoricos Gente, / In Galliâ natus . . ."; the funeral oration was by the Leiden historiographer Marcus Zuerius Boxhornius.

3. The family's coat of arms was awarded following a determination of the family's nobility in 1668: C. Adam, *Vie et Oeuvres*, 13.

4. For a helpful family tree, see D. Clarke, *Descartes: A Biography*, x–xi.

5. C. Adam, *Vie et Oeuvres*, 586. He died after an illness of several days, in the home of the French ambassador, Chanut, who knew Descartes well, so this burial location was presumably in accordance with Descartes's wishes.

6. D'Argenson, "Note sur la famille Descartes," 87–88, 92, 95.

7. James Michael Hayden, *France and the Estates General of 1614* (Cambridge: Cambridge University Press, 1974), 87.

8. C. Adam, *Vie et Oeuvres*, 1–18; Watson, *Cogito*, 44; Rodis-Lewis, *Descartes: Life and Thought*, 2.

9. The negative result is from *Dictionnaire de biographie Française* (Paris: Letouzey et Ané, 1965), 10:1238–44; the ambassador was Emeric Gobier, Sieur de Barrault.

10. Pierre Mathurin de l'Écluse des Loges, *Memoirs of Maximilian de Bethune, Duke of Sully*, trans. Charlotte Lennox and Samuel Johnson, 6 vols. (London: J. Rivington et al. 1778), 4:25–34; Alexandre Petitot et al., *Collection des memoires relatifs a l'histoire de France* (Paris: Foucault, 1824), 44:467.

11. On April 10, 1618: C. Adam, *Vie et Oeuvres*, 40.

12. The letters were dated December 10, 1625, which would refer back to Joachim's forty years in office in the *parlement* (C. Adam, *Vie et Oeuvres*, 11).

13. Hayden, *France and Estates General*, 79.

14. Leroy, *Descartes: Philosophe au Masque*, 47.

15. C. Adam, *Vie et Oeuvres*, 7–8. Descartes's mixed but ultimately negative view of Machiavelli can be found in a letter to Princess Elizabeth immediately after she was ordered to her relatives in Brandenburg, September 1646 (Cottingham, Stoothoff, and Murdoch, *Philosophical Writings*, 3:292–95).

16. For a fine study of Jesuit education and La Fleche, see Gaukroger, *Descartes: Intellectual Biography*, 38–61. Descartes studied there from about Easter 1606 probably until 1613 or 1614 (the evidence is unclear). Gaukroger (*Descartes: Intellectual Biography*, 38), and Watson (*Cogito*, 76) think that the 1614 date is more likely; Rodis-Lewis (*Descartes: Life and Thought*, 8) prefers September 1615.

17. C. Adam, *Vie et Oeuvres*, 20, 32–33.

18. C. Adam and Tannery, *Oeuvres De Descartes*, 10:535.

19. Rodis-Lewis, *Descartes: Life and Thought*, 21–22.

20. C. Adam, *Vie et Oeuvres*, 11, 13.

21. Ibid., 36–37.

22. Hayden, *France and Estates General*, 72, 79–80.

23. On the dispute in Poitiers, see Jeffrey K. Sawyer, *Printed Poison: Pamphlet Propaganda, Faction Politics, and the Public Sphere in Early Seventeenth-Century France* (Berkeley: University of California Press, 1990), 73–83.

24. Hayden, *France and Estates General*, 280; D. Clarke, *Descartes: A Biography*, xi. *Maître* is a French form of address for senior lawyers.

25. Hayden, *France and Estates General*, 79, 283.

26. Arlette Jouanna, *The St. Bartholomew's Day Massacre: The Mysteries of a Crime of State* (Manchester, UK: Manchester University Press, 2013), 233.

27. Hilary Gatti, *Ideas of Liberty in Early Modern Europe: From Machiavelli to Milton* (Princeton, NJ: Princeton University Press, 2015), 117–33.

28. Hayden, *France and Estates General*, 131–4, 145.

29. John Cottingham, ed., *Descartes' Conversation with Burman*, (Oxford: Clarendon Press, 1976), 33.

30. For instance, C. Adam, *Vie et Oeuvres*, 28; also see Toulmin, *Cosmopolis*, 56–62.

31. D'Argenson, "Note sur la famille Descartes," 91.

32. For a description of the entombment of Henri's heart, see Christian Regniér, "The Heart of the Kings of France: 'Cordial Immortality,'" Medicographia, accessed April 17, 2017, http://www.medicographia.com/2010/07/the-heart-of-the-kings-of-france-cordial-immortality/; on Descartes's burial and tomb, see C. Adam, *Vie et Oeuvres*, 585–94.

33. For example, C. Adam, *Vie et Oeuvres*, 69.

34. Rodis-Lewis, *Descartes: Life and Thought*, 23.

35. C. Adam, *Vie et Oeuvres*, 69; G. Cohen, *Écrivains Français en Hollande*, 413; Watson, *Cogito*, 123; Rodis-Lewis, *Descartes: Life and Thought*, 61.

36. Baillet, *Des-Cartes*, 1:136. Baillet is not entirely clear about the date but says that it was about a year after Descartes's return from Italy, which would place it in the spring or early summer of 1626.

37. *Mémoires du Cardinal de Richelieu, tome sixième (1626)* (Paris: Édouard Champion, 1925), 74–75.

38. Hayden, *France and Estates General*, 249.

39. Ibid., 159, 53.

40. Kleinman, *Anne of Austria*, 33; Carl Jacob Burckhardt, *Richelieu and His Age*, trans. Edwin Muir, Willa Muir, and Bernard Hoy, 3 vols. (London: Allen and Unwin, 1967), 1:150–57.

41. Jean-Marie Constant, *Gaston d'Orleans: Prince de la Liberté* (Paris: Perrin, 2013), 116: "'maintien de la liberté du peuple, sans blesser l'autorité du prince.'" The oration was pronounced by Jean François Senault, who wrote on the passions and become head of the Oratorians in 1663.

42. Descartes quoted in Cottingham, Stoothoff, and Murdoch, *Philosophical Writings*, 1:384–5.

43. The story has often been told. I have mainly followed the versions in Batiffol, *Duchesse de Chevreuse*, 86–117, and Burckhardt, *Richelieu and His Age*, 192–207.

44. *Mémoires du Cardinal de Richelieu*, 6:74–75: Jean de Bourgneuf, sieur de Cucé, and Isaac Loisel de Bry were the premier and second presidents, respectively; Descartes served as doyen; Hay as sous-doyen, with the remaining councilors being Gilles de Lys, Laurent Peschart, Jean du Halgouët, de Martigné, Oudart, Huet, and François d'Andigné; the clerks were Pierre Malescot and Pierre de Verdun. François Foucquet, father of the celebrated *surintendant*, Charles de Machault, siegneur d'Arnouville, and Tanneguy de Launay acted for the prosecution.

45. Kleinman, *Anne of Austria*, 69–70.

46. C. Adam, *Vie et Oeuvres*, 11.

47. The award was given "in the camp before La Rochelle, 20 July 1628" (ibid., 11).

48. Cottingham, Stoothoff, and Murdoch, *Philosophical Writings*, 1:115.

49. On Descartes's education, see esp. Gaukroger, *Descartes: Intellectual Biography*, 38–61.

50. C. Adam and Tannery, *Oeuvres De Descartes*, 10:535.

51. "Son devoir joint à son inclination le portoit à vouloir prendre parti dans les troupes du roy: mais il fallut prendre quelques mesures pour ne point paroître partisan du maréchal D'Ancre" ("His duty and inclination together induced him to want to join with the king's company: but it was necessary to take measures for not appearing to be a partisan of the maréchal D'Ancre," i.e., Concini): Baillet, *Des-Cartes*, 1:40.

52. Baillet, *Des-Cartes*, 1:35–39. He must here be thinking of the possibility that Descartes accompanied the court to Guyenne, since the proxy marriages both occurred on October 18, 1615, and all the parties met on November 9; the wedding in France was celebrated on November 25 in Bordeaux, and the return to Paris began at the end of November, taking four months. The court's return passed through Poitiers, as well as Tours and Blois: Kleinman, *Anne of Austria*, 23–29.

53. Baillet, *Life*, 22.

54. The literature on Renaissance manners is extensive, much stimulated by Norbert Elias, *The Civilizing Process*, trans. Edmund Jephcott, 2 vols. (New York: Urizen, 1978–1982); for a fine recent introduction to the genre, see Snyder, *Dissimulation*. By Descartes' lifetime "masking" had shifted from aesthetic to political purposes: Marc Bayard, "Double *Persona*: Le Masque, D'une esthéthique à une politique (XVIe–XVIIe Siècles)," in *Masques Mascarades Mascarons*, edited by Françoise Viatte, Dominique Cordellier, and Violaine Jeammet (Paris: Musée du Louvre, 2014), 181–89.

55. Cottingham, Stoothoff, and Murdoch, *Philosophical Writings*, 1:367–69.

56. Jacques Thuillier and Jacques Foucart, *Rubens' Life of Marie De Medici* (New York: Harry N. Abrams, [1970]), 27.

57. Kleinman, *Anne of Austria*, 31.

58. Kathleen Wellman, *Queens and Mistresses of Renaissance France* (New Haven, CT: Yale University Press, 2013), 245.

59. Bassompierre, *Memoires du Mareschal de Bassompierre, contenant l'histoire de sa vie: Et de ce qui s'est fait de plus remarquable à la cour de France pendant quelques années* (1665; Cologne: Jean Sambix, 1703), 78–79.

60. Trevor Aston, ed. *Crisis in Europe, 1560–1660: Essays from Past and Present* (London: Routledge and Kegan Paul, 1970); Geoffrey Parker, *Global Crisis: War, Climate Change and Catastrophe in the Seventeenth Century* (New Haven, CT: Yale University Press, 2013).

61. Geoffrey Keynes, *The Life of William Harvey* (Oxford: Clarendon Press, 1966), 194.

62. Baillet, *Des-Cartes*, 1:36–37.

63. F. E. Sutcliffe, *Guez de Balzac et son temps: Littérature et politique* (Paris: A. G. Nizet, 1959).

64. Ibid., 18, 23. The duc d'Épernon is usually seen as one of Marie de Medici's closest allies. For his involvement in the Chalais plot, see Victor-L. Tapié, *France in the Age of Louis XIII and Richelieu*, trans. D. McN. Lockie (London: Macmillan, 1974), 161.

65. The phrase was made well-known by Pintard, *Libertinage Érudite*.

66. La Mothe Le Vayer married Mademoiselle de la Haye, daughter of the ambassador, i.e., the niece of Mydorge's wife: David Durand, *The Life of Lucilio (Alias Julius Caesar) Vanini, Burnt for Atheism at Thoulouse* (London: Printed for W. Meadows at the Angel in Cornhill, 1730), 93. Mydorge died in 1647, when she was said to be about forty-two; La Mothe Le Vayer died in 1672.

67. Harth, *Cartesian Women*, 15–16.

68. On Gournay and Théophile, see Harth, *Cartesian Women*, 29; on Gournay and Balzac, see Marjorie Henry Ilsley, *A Daughter of the Renaissance: Marie le Jars de Gournay, Her Life and Works* (The Hague: Mouton, 1963), 134–35, 226; Gournay frequented the libertine circle around Marguerite de Valois: Jacqueline Broad and Karen Green, *A History of Women's Political Thought in Europe, 1400–1700* (Cambridge: Cambridge University Press, 2009), 135.

69. Richard H. Popkin, *The History of Scepticism from Erasmus to Descartes* (New York: Harper and Row, 1968); Cottingham, Stoothoff, and Murdoch, *Philosophical Writings*, 1:119; C. Adam and Tannery, *Oeuvres De Descartes*, 6:16; Montaigne's essay on cannibals is well known; on Gournay's interest in the emperors of China, see Broad and Green, *History of Women's Political Thought*, 128.

70. Jason Lewis Saunders, *Justus Lipsius: The Philosophy of Renaissance Stoicism* (New York: Liberal Arts Press, 1955).

71. Rodis-Lewis, *Descartes: Life and Thought*, 27; Cottingham, Stoothoff, and Murdoch, *Philosophical Writings*, 1:347. For Descartes's friend Balzac denounc-

ing the immoral consequences of Stoic paradoxes, see Roger Zuber, ed., *Oeuvres Diverses (1644)* (Paris: Honoré Chamion Éditeur, 1995), 128.

72. Stephen Greenblatt, *The Swerve: How the World Became Modern* (New York, London: W. W. Norton, 2011); Ada Palmer, *Reading Lucretius in the Renaissance* (Cambridge, MA: Harvard University Press, 2014); C. Wilson, *Epicureanism*; Snyder, *Dissimulation*.

73. Cremonini's private motto was *Intus ut libet, foris ut moris est* ("In private think what you like, in public behave as is the custom"): John S. Spink, *French Free-Thought from Gassendi to Voltaire* (London: Athlone, 1960), 9.

74. See especially C. Wilson, *Epicureanism*.

75. Pierre Bayle, *A General Dictionary, Historical and Critical*, 10 vols. (London: James Bettenham et al., 1734–1741), 4:460–61.

76. Johann Weyer, *Witches, Devils, and Doctors in the Renaissance: Johann Weyer, De Praestigiis Daemonum*, trans. John Shea (Binghamton, NY: Medieval and Renaissance Texts and Studies, 1991), 529–35.

77. Marijke Gijswijt-Hofstra and Willem Frijhoff, eds. *Witchcraft in the Netherlands from the Fourteenth to the Twentieth Century* (Rotterdam: Universtaire Pers, 1991).

78. Robin, Briggs, *The Witches of Lorraine* (Oxford: Oxford University Press, 2007), 20–21.

79. Quotations from Cottingham, Stoothoff, and Murdoch, *Philosophical Writings*, 2:15, 62, 61, 11.

80. The literature on witchcraft is enormous, but for a line about its associations with changes in legal procedure see, for example, Brian P. Levack, *The Witch-Hunt in Early Modern Europe* (London: Longman, 1987).

81. Nicholas S. Davidson, "Lucretius, Atheism, and Irreligion in Renaissance and Early Modern Venice," in *Lucretius and the Early Modern*, ed. David Norbrook, S. J. Harrison, and Philip R. Hardie (Oxford: Oxford University Press, 2016), 67–79.

82. Leopizzi, *Les sources documentaires*, 732–33; for a more contemporary reading, see Durand, *Life of Vanini*; Foucault, *Philosophe libertin*; Cavaillé, *Dis/simulations*; Charles-Daubert, *Libertins Érudits*, 25–26; Georges Minois, *The Atheist's Bible: The Most Dangerous Book That Never Existed* (Chicago: University of Chicago Press, 2012), 90–91.

83. Minois, *Atheist's Bible*, 58, quoting the Jesuit Garasse.

84. Germana Ernst, *Tommaso Campanella: The Book and the Body of Nature*, trans. David L. Marshall (Dordrecht; New York: Springer, 2010), 34.

85. Bertram Eugene Schwarzbach and A. W. Fairbairn, "History and Structure of Our Traité des Trois Imposteurs," in *Heterodoxy, Spinozism, and Free Thought in Early-Eighteenth-Century Europe: Studies on the Traité des Trois Imposteurs*, ed. Silvia Berti, Françoise Charles-Daubert, and Richard H. Popkin (Dordrecht; Boston: Kluwer, 1996), 90–91; for the story of the "three impostors," see Minois, *Atheist's*

Bible, and the essays in Berti, Charles-Daubert, and Popkin, *Heterodoxy, Spinozism, and Free Thought*.

86. Anne Staquet, *Descartes et le Libertinage*, 11–17; Snyder, *Dissimulation*.

87. Pintard, *Libertinage Érudit*; Charles-Daubert, *Libertins Érudits*.

88. Peter Dear, *Mersenne and the Learning of the Schools* (Ithaca, NY: Cornell University Press, 1988), 23–27.

89. Durand, *Life of Vanini*, 101.

90. For a summary of his life and work, see W. D. Howarth, *Life and Letters in France: The 17th Century* (New York: Charles Scbribner's, 1965), 14–19.

91. Sutcliffe, *Guez de Balzac*, 27.

92. Balzac's relationship with Théophile later cooled for an unknown reason: ibid., 22.

93. Adam, *Vie et Oeuvres*, 412, 416.

94. Batiffol, *Duchesse de Chevreuse*, 17.

95. René Descartes to Pierre Chanut, 1 February 1647, in Cottingham, Stoothoff, and Murdoch, *Philosophical Writings*, 3:313, Descartes quotes Théophile on the power of love to destroy, as Paris's desire did to Troy.

96. "Son devoir joint à son inclination le portoit à vouloir prendre parti dans les troupes du roy: mais il fallut prendre quelques mesures pour ne point paroître partisan du maréchal D'Ancre.": Baillet, *Des-Cartes*, 1:40.

97. Thuillier and Foucart, *Rubens' Life of Marie De Medici*, 17.

98. Baillet, *Des-Cartes*, 131.

99. Kleinman, *Anne of Austria*, 20; Constant, *Gaston d'Orleans*, 45–46, 201–35.

100. C. Adam, *Vie et Oeuvres*, 590.

101. Shorto, *Descartes' Bones*, 43–53, 65–70. It is perhaps similar to Galileo's (middle) finger, now on exhibit in the Galileo museum in Florence.

102. Baillet, *Des-Cartes*, 1:39.

Part Three

1. Among other studies, see William Eamon, *Science and the Secrets of Nature: Books of Secrets in Medieval and Early Modern Culture* (Princeton, NJ: Princeton University Press, 1994); Pamela O. Long, *Openness, Secrecy, Authorship: Technical Arts and the Culture of Knowledge from Antiquity to the Renaissance* (Baltimore: Johns Hopkins University Press, 2001); Pamela Smith, *The Body of the Artisan: Art and Experience in the Scientific Revolution* (Chicago: University of Chicago Press, 2004); Pamela H. Smith, Amy R. W. Meyers, and Harold J. Cook, eds., *Ways of Making and Knowing: The Material Culture of Empirical Knowledge* (Ann Arbor: University of Michigan Press, 2014).

2. Baillet, *Des-Cartes*, 1:40: "En quoi il se proposa l'éxemple de plusieurs juenes gentilshommes de la noblesse françoise, qui alloient alors apprendre le métier de la guerre sous le prince Maurits De Nassau en Hollande."

3. Rodis-Lewis, *Descartes: Life and Thought*, 22.

4. Cottingham, Stoothoff, and Murdoch, *Philosophical Writings*, 3:4–5.

5. Maurits's brother had been raised by the Spaniards, brought up a Catholic, and married into the French nobility (Eleonora of Bourdon-Condé). He also inherited the barony of Breda, so that Philip William would have been its governor, not Maurits, if Descartes had gone there in 1617. Baillet was aware of these distinctions when he published the biography of Descartes in 1691 (Baillet, *Des-Cartes*, 1:44–49), and he elaborated on them in a book on the Princes of Orange in 1692, which includes a section on Philip William (pp. 75–89) and well as Maurits (pp. 91–109): Adrien Baillet, *Histoire des Princes d'Orange de la Maison de Nassau* (Amsterdam: Paul Marret, 1692).

6. C. Adam, *Vie et Oeuvres*, 42–43.

7. François Monnier, *Philippe de Béthune (1565–1649): Le conseiller d'estat, ou, recueil général de la politique moderne* (Paris: Economica, 2012).

8. Isaac Beeckman, *Journal tenu par Isaac Beeckman de 1604 à 1634*, 4 vols., ed. Cornelis de Waard (La Haye: Martinus Nijhoff, 1939–53), 237: "Nitebatur heri, qui erat 10 Nov. 1618; Bredae Gallus Picto probare nullum esse angulum revera, hoc argumento: . . ."; and 257, entry for December 26, 1618.

9. Adam, *Vie et Oeuvres*, 42, accepts Borel's attestation on this, as does Rodis-Lewis, *Descartes: Life and Thought*, 25. As far as I know, none of his other biographers have made anything of the fact that the coin was a doubloon. Borel calls it a "duplio," referring to the Spanish *doblón*, meaning "double": Borel, *Renatus Descartes*, 8.

10. René Descartes, *Compendium of Music (Compendium Musicae)*, ed. Charles Kent, trans. Walter Robert (Münster, Germany: American Institute of Musicology, 1961), conclusion.

11. René Descartes to Servien, 12 May 1647, in C. Adam and Tannery, *Oeuvres de Descartes*, 5:25, cited by G. Cohen, *Écrivains Français en Hollande*, 374.

12. Baillet, *Des-Cartes*, 1:44–45.

13. Cottingham, *Descartes' Conversation*, sec. 51, p. 32; for the sake of simplicity, I have substituted "Counter-Remonstrant" for his term, "Gomarist," and "Remonstrant" for "Arminian." After Descartes's death, a note from the French embassy mentioned that he was *Prédestiné*: C. Adam, *Vie et Oeuvres*, 586.

14. Frederick B. Artz, *The Development of Technical Education in France, 1500–1850* (Cambridge, MA: MIT Press, 1966), 42; Gaukroger, *Descartes: Intellectual Biography*, 59.

15. Henry Guerlac, "Science and War in the Old Regime: The Development of Science in an Armed Society" (PhD diss., Harvard University, 1941), 63–85.

16. Ken Alder, *Engineering the Revolution: Arms and Enlightenment in France, 1763–1815* (Princeton, NJ: Princeton University Press, 1997).

17. Eric Lund, "The Generation of 1683: The Scientific Revolution and Generalship in the Habsburg Army, 1686–1723," in *Warfare in Eastern Europe, 1500–1800*, ed. Brian L. Davies (Boston: Brill, 2012), 199–248.

18. An excellent recent study is Marjolein 't Hart, *The Dutch Wars of Independence: Warfare and Commerce in the Netherlands 1570–1680* (London: Routledge, 2014).

19. For a fine general account, see James A. Bennett, "The Mechanical Arts," in *The Cambridge History of Science*, ed. Katherine Park and Loraine Daston (Cambridge: Cambridge University Press, 2006), 686–93.

20. See esp. Tonio Andrade, *The Gunpowder Age: China, Military Innovation, and the Rise of the West in World History* (Princeton, NJ: Princeton University Press, 2016), 1–23.

21. A compelling overview is Lauro Martines, *Furies: War in Europe, 1450–1700* (New York: Bloomsbury Press, 2013); see also Christopher Duffy, *Siege Warfare: The Fortress in the Early Modern World, 1494–1660* (London: Routledge and Kegan Paul, 1979).

22. The words are from Buonaiuto Lorini, quoted in Alexander Marr, *Between Raphael and Galileo: Mutio Oddi and the Mathematical Culture of Late Renaissance Italy* (Chicago: University of Chicago Press, 2011), 81.

23. Pamela O. Long, David McGee, and Alan M. Stahl, eds., *The Book of Michael of Rhodes: A Fifteenth-Century Maritime Manuscript*, 3 vols. (Cambridge, MA.: MIT Press, 2009).

24. Luigi Barbasetti, *The Art of Foil, with a Short History of Fencing* (New York: E. P. Dutton, 1932), 175–274; Evelyn Lincoln, *Brilliant Discourse: Pictures and Readers in Early Modern Rome* (New Haven, CT: Yale University Press, 2014), 82–102. The most notable example of geometrical design for fencing is Girard Thibault, *Academie de l'espée de Girard Thibault d'Anvers: Ou se demonstrent par reigles mathematiques sur le fondement d'un cercle mysterieux la theorie et pratique des vrais et iusqu'a present incognus secreta du maniement des armes, a pied et a cheval* (Anvers [Antwerp]: 1628).

25. G. Cohen, *Écrivains Français en Hollande*, 373; the French engineers were Jacques Alleaume and David van Orliens.

26. Fokko Jan Dijksterhuis, "Geometries of Space: Dutch Mathematics and the Visualization of Distance," in *Mapping Spaces: Networks of Knowledge in 17th Century Landscape Painting*, ed. Ulrike Gehring and Peter Weibel (Karlsruhe, Germany: Hirmer, 2014), 349.

27. Pieter Jan van Winter, *Hoger beroepsonderwijs avant-la-lettre: Bemoeiingen met de vorming van landmeters en ingenieurs bij de Nederlandse Universiteiten van de 17ᵉ en 18ᵉ eeuw* (Amsterdam: Noord-Hollandsche Uitg. Mij, 1988), 16–22.

28. E. J. Dijksterhuis, *Simon Stevin* (The Hague: Martinus Nijhoff, 1943); Dirk J. Struik, *The Land of Stevin and Huygens: A Sketch of Science and Technology in the Dutch Republic during the Golden Century*, Studies in the History of Modern Science, No. 7 (1958; Dordrecht: D. Reidel, 1981); Klaas van Berkel, "Part One: The Legacy of Stevin: A Chronological Narrative," in *A History of Science in the Netherlands: Survey, Themes, and Reference*, ed. Klaas van Berkel, Albert van Helden, and Lodewijk Palm (Leiden: E. J. Brill, 1999).

29. Michael Roberts, *The Military Revolution 1560–1660* (Belfast: M. Boyd, 1956); Geoffrey Parker, *The Military Revolution: Military Innovation and the Rise of the West, 1500–1800* (Cambridge: Cambridge University Press, 1996); 't Hart, *Dutch Wars of Independence*.

30. 't Hart, *Dutch Wars of Independence*, 37–80.

31. That he knew how to describe simple machines is clear from his little work of October 1637, sent as a letter to Huygens; Cottingham, Stoothoff, and Murdoch, *Philosophical Writings*, 3:66–73.

32. My translation, based on Cottingham, Stoothoff, and Murdoch, *Philosophical Writings*, 1:116–17, but restoring the literal meaning based on the 1637 edition, and italicizing two words.

33. An alternative reading would be to imagine he was thinking of the Ville de Richelieu, on which work began in 1631, but that was more a grand château than a town per se. By the time of the *Discours* he would have passed by Glückstadt.

34. My translation, based on Cottingham, Stoothoff, and Murdoch, *Philosophical Writings*, 1:116–17, but restoring the literal meaning based on the 1637 edition, and italicizing one word.

35. Klaas van Berkel, *Isaac Beeckman on Matter and Motion: Mechanical Philosophy in the Making* (Baltimore: Johns Hopkins University Press, 2013), 147.

36. Beeckman, *Journal*, 4:62, also quoted in Berkel, *Isaac Beeckman*, 26.

37. Baillet, *Des-Cartes*, 1:xiii–xiv. Van Schooten also edited Viète's works in 1646. An English translation of the passage is given in Baillet, *Life*, 23–24. Lipstorp published an account of Descartes's life as an appendix to his 1653 work on Descartes's ideas: C. Adam, *Vie et Oeuvres*, v–vi. The work is summarized in Borel, *Renatus Descartes*, 65–107. Lipstorp had later taken up residence in Paris and become well acquainted with Baillet, giving him everything he had on Descartes (as well as a work on how to decode the scriptures to predict the future): Baillet, *Jugemens des Savans*, 19.

38. On mathematical tutors in the period, see James A. Bennett, "The Mechanics' Philosophy and the Mechanical Philosophy," *History of Science* 24 (1986): 1–28; Mario Biagioli, "The Social Status of Italian Mathematicians, 1450–1600," *History of Science* 27 (1989): 41–95; Fokko Jan Dijksterhuis, "Stevin, Huygens and the Dutch Republic: The Golden Age of Mathematics," *Nieuw archief voor wiskunde*, 5th ser., 9 (2008): 100–107.

39. The mother of Benjamin de Rohan, baron de Soubise, an important Huguenot military leader who will figure in the subsequent account, and who had also studied war under Maurice of Nassau.

40. Adam, *Vie et Oeuvres*, 215. See also *Wikipedia*, s.v. "François Viète," last modified April 7, 2017, https://en.wikipedia.org/wiki/François_Viète.

41. For instance, the Venetian Paolo Sarpi thought Viète was one of two original minds from his own century, the other being Humphrey Gilbert: Massimo Bucciantini, Michele Camerota, and Franco Giudice, *Galileo's Telescope: A European*

Story, trans. Catherine Bolton (Cambridge, MA: Harvard University Press, 2015), 35.

42. J. P. Devos, *Les Chiffres de Philippe II (1555–1598) et du despacho universal durant le XVIIe siècle* (Brussels: Palais des Académies, 1950), 59; David Kahn, *The Codebreakers: The Story of Secret Writing* (1967; New York: Scribner, 1996), 116–18; Peter Pesic, "Secrets, Symbols, and Systems: Parallels between Cryptanalysis and Algebra," *Isis* 88 (1997): 674–92; Dejanirah Couto, "Spying in the Ottoman Empire: Sixteenth-Century Encrypted Correspondence," in *Cultural Exchange in Early Modern Europe*, vol. 3, ed. Francisco Bethencourt and Florike Egmond (Cambridge: Cambridge University Press, 2007); Kristie Macrakis, "Confessing Secrets: Secret Communication and the Origins of Modern Science," *Intelligence and National Security* 25 (2010): 183–97.

43. Berkel, *Isaac Beeckman*, 22.

44. On his departure and unspecified travels in the province of Holland, see ibid., 25.

45. Ibid., 29.

46. Ibid., 14–15: his first private teacher was a distant relative, Jan van den Broecke.

47. "Mr. Duperon Picto Renatus Descartes vocatur in ea Musica, quam mea causa jam describit": Beeckman, *Journal*, 257.

48. Cottingham, Stoothoff, and Murdoch, *Philosophical Writings*, 3:4; Berkel, *Isaac Beeckman*, 25.

49. Berkel, *Isaac Beeckman*, 19.

50. H. H. Kubbinga, "Beeckmans 'Molecuul'-Theorie als Nieuwe Categorie in de Geschiedenis van de Theorie van de Materie: Een Overzicht," *Algemeen Nederlands Tijdschrift voor Wijsbegeerte* 81 (1989): 161–77.

51. An examination of Beeckman's notebook shows that music probably constituted the bulk of his journal writing in the period, and that those sections are not so much summaries of others' opinions (as with his medical notes) as they are investigations exploring his own ideas: Beeckman, *Journal*, vol. 1.

52. On music and natural philosophy, see H. F. Cohen, *Quantifying Music: The Science of Music At the First Stage of the Scientific Revolution, 1580–1650* (Dordrecht: Reidel, 1984); D. P. Walker, *Music, Spirit and Language in the Renaissance*, ed. Penelope Gouk (London: Variorum Reprints, 1985); Penelope Gouk, *Music, Science and Natural Magic in Seventeenth-Century England* (New Haven, CT: Yale University Press, 1999).

53. Descartes, *Compendium of Music*.

54. Berkel, *Isaac Beeckman*, 25.

55. Schuster, *Descartes-Agonistes*, 112–28.

56. Baillet, *Des-Cartes*, 1:50–51: the work was called *De l'Ame des Bêtes*.

57. Descartes, *Compendium of Music*, conclusion.

58. C. Adam and Tannery, *Oeuvres De Descartes*, 10:151–2; G. Cohen, *Écrivains*

Français en Hollande, 381–82; the word *pictura* is often translated as "painting" (as in Cottingham, Stoothoff, and Murdoch, *Philosophical Writings*, 3:1), but I think in this context it implies three-dimensional perspective drawing.

59. Cottingham, *Descartes' Conversation*, 47.

60. Mary J. Henninger-Voss, "How the 'New Science' of Cannons Shook Up the Aristotelian Cosmos," *Journal of the History of Ideas* 63 (2002): 371–97; Mary J. Henninger-Voss, "Comets and Cannonballs: Reading Technology in a Sixteenth-Century Library," in *The Mindful Hand: Inquiry and Invention from the Late Renaissance to Early Industrialisation*, ed. Lissa Roberts, Simon Schaffer, and Peter Dear (Amsterdam: Koninklijke Nederlandse Akademie van Wetenschappen, 2007).

61. Pamela O. Long, *Artisan/Practitioners and the Rise of the New Sciences, 1400–1600* (Corvalis: Oregon State University Press, 2011), 106–7; Martines, *Furies*, 63–64.

62. Matteo Valleriani, *Galileo Engineer* (Dordrecht: Springer, 2010); see also Mario Biagioli, *Galileo's Instruments of Credit: Telescopes, Images, Secrecy* (Chicago: University of Chicago Press, 2006); J. L. Heilbron, *Galileo* (Oxford: Oxford University Press, 2010).

63. Sven Dupré, "Ausonio's Mirrors and Galileo's Lenses: The Telescope and Sixteenth-Century Practical Optical Knowledge," *Galilaeana* 2 (2005): 145–80.

64. Bucciantini, Camerota, and Giudice, *Galileo's Telescope*; for a recent account of the probable development of the telescope in the Dutch Republic, see Arjen Dijkstra, "Between Academics and Idiots: A Cultural History of Mathematics in the Dutch Province of Friesland (1600–1700)" (PhD diss., Twente University, 2012), 132–56.

65. Ofer Gal and Raz D. Chen-Morris, *Baroque Science* (Chicago: University of Chicago Press, 2013).

66. See especially Marr, *Between Raphael and Galileo*; Ulrike Gehring and Peter Weibel, eds., *Mapping Spaces: Networks of Knowledge in 17th Century Landscape Painting* (Karlsruhe, Germany: Hirmer, 2014).

67. René Descartes to Isaac Beeckman, 26 March 1618, in Cottingham, Stoothoff, and Murdoch, *Philosophical Writings*, 3:1–4; see also his disappointment in realizing that it was not so simple to make the devices he imagined (p. 4).

68. Dijksterhuis, "Geometries of Space"; the work referred to is Van Schooten's *Organica conicarum sectionum in plano descriptione* ("Mechanical description of conic sections in a plane," 1646).

69. H. J. M. Bos, ed., *The Structure of Descartes' Géometrie*, Lectures in the History of Mathematics (Providence, RI: American Mathematical Society, 1993).

70. Dijksterhuis, "Geormetries of Space," 115.

71. Cottingham, Stoothoff, and Murdoch, *Philosophical Writings*, 3:1–4.

72. The "Parnassus" notebook is inscribed "1 Jan 1619": C. Adam, *Vie et Oeuvres*, 29.

73. René Descartes to Isaac Beeckman, 24 January 1619, in Cottingham, Stoothoff, and Murdoch, *Philosophical Writings*, 3:1.

74. The letter was sent from Breda to "Monsieur Isaac Beeckman Docteur en medicine" at his parents' house on the Beestenmarkt in Middleburg (C. Adam and Tannery, *Oeuvres De Descartes*, 10:160).

75. Berkel, *Isaac Beeckman*, 26.

76. René Descartes to Isaac Beeckman, 23 April 1618, in Cottingham, Stoothoff, and Murdoch, *Philosophical Writings*, 3:3.

77. René Descartes to Isaac Beeckman, 29 April 1618, in Cottingham, Stoothoff, and Murdoch, *Philosophical Writings*, 3:4.

78. Isaac Beeckman to René Descartes, 6 May 1618, in C. Adam and Tannery, *Oeuvres De Descartes*, 10:167–69.

79. Baillet, *Des-Cartes*, 1:53.

80. Leroy, *Descartes: Philosophe au Masque*, 155–56. What little is known of Villebressieu is published in C. Adam and Tannery, *Oeuvres De Descartes*, 1:214–15, 218. He may have been a member of the noble family of Bressieux associated with the dukes of Burgundy, whose castle was destroyed in 1612.

81. Opening sentence of part 2 of *Discours*.

82. Baillet, *Des-Cartes*, 1:54.

83. Antoine Adam, *Théophile de Viau et la libre pensée Française en 1620* (1935; repr., Geneva: Slatkine Reprints, 1965), 433.

84. On Postel, see William J. Bouwsma, *Concordia Mundi: The Career and Thought of Guillaume Postel (1510–1581)* (Cambridge, MA: Harvard University Press, 1957); Susanna Åkerman, *Rose Cross over the Baltic: The Spread of Rosicrucianism in Northern Europe* (Leiden: Brill, 1998), 173–95; Jean Bodin, *Colloquium of the Seven about Secrets of the Sublime: Colloquium Heptaplomeres de Rerum Sublimium Arcanis Abditis*, trans. Marion Leathers Daniels Kuntz (Princeton, NJ: Princeton University Press, 1975); Ann Blair, *The Theater of Nature: Jean Bodin and Renaissance Science* (Princeton, NJ: Princeton University Press, 1997).

85. Richard H. Popkin, *Isaac La Peyrère (1596–1676): His Life, Work, and Influence* (Leiden: E. J. Brill, 1987), 7, 12–13.

86. Benoist Pierre, *Le Père Joseph: L'Eminence Grise de Richelieu* (Paris: Perrin, 2007), 225.

87. Jonathan Spangler, *The Society of Princes: The Lorraine-Guise and the Conservation of Power and Wealth in Seventeenth-Century France* (Farnham, UK: Ashgate, 2009), 199, 201.

88. Susanna Åkerman, "Queen Christina of Sweden and Messianic Thought," in *Sceptics, Millenarians and Jews*, ed. David S. Katz and Jonathan I. Israel (Leiden: Brill, 1990), 155–60; Åkerman, *Queen Christina*, 11–13, 230–33.

89. Åkerman, *Queen Christina*, 13, 55.

90. Baillet, *Des-Cartes*, 1:55.

91. Monnier, *Philippe de Béthune*, 43–46.

92. Cottingham, Stoothoff, and Murdoch, *Philosophical Writings*, 1:81.

93. Baillet, *Des-Cartes*, 1:55–58.

94. Ibid.

95. "It se mit donc dans les troupes bavaroises comme simple volontaire sans voulour prendre d'employ." Ibid., 58.

96. Burckhardt, *Richelieu and His Age*, 171.

97. Spangler, *Society of Princes*. For instance, Charles III, Duke of Lorraine, married Claude de Valois, daughter of Henri and Catherine de Medici, in 1559; his son, Henry II "the good," who had first married the sister of Henri IV, remarried in 1606 to Margherita Gonzaga, Marie de Medici's niece; their granddaughter, Margaret, in turn secretly married Gaston d'Orleans in 1632. While Lorraine would be occupied by France under Richelieu, the Peace of Westphalia in 1648 forced France's withdrawal.

98. Cole, *Olympian Dreams*. The date is consistent with that in Descartes's surviving notebook, the "Cogitationes Privatae," in C. Adam and Tannery, *Oeuvres De Descartes*, 10:216. The vivid and detailed description of the dreams might have been a draft of a book. Although they have long been taken as "real," they are revealing even if read metaphorically.

99. Quoted in Cottingham, Stoothoff, and Murdoch, *Philosophical Writings*, 1:116.

100. I follow the translation in ibid., 32–40.

101. Cottingham, Stoothoff, and Murdoch, *Philosophical Writings*, 1:117–25. His informal moral rules (rather than absolute ones) were entirely in keeping with contemporary aristocratic norms: Emma Gilby, "Descartes's 'Morale Par Provision': A Re-Evaluation," *French Studies* 65 (2011): 444–58.

102. Cole, *Olympian Dreams*, 32–40.

103. Robert Halleux, "Helmontiana II: Le Prologue de l'*Eisagoge*, la Conversion de Van Helmont au Paracelsisme et les Songes de Descartes," *Mededelingen van de koninklijke academie voor wetenschappen, letteren en schone kunten van België, Klasse der Wetenschappen* 49, no. 2 (1987): 32–33.

104. Ibid.

105. Frédéric de Buzon, "Un exemplaire de la Sagesse de Pierre Charron offert à Descartes en 1619," *Archives de Philosophie* 55, no. 20 (1992): 1–3, with the Latin original: "Doctissimo Amico grato et minori fratri Renato Cartesio, d.d. ded., P. Johannes B. Molitor S.J., exeunte Anno 1619, JBM."

106. Tullio Gregory, "Pierre Charron's 'Scandalous Book'," in *Atheism from the Reformation to the Enlightenment*, ed. Michael Hunter and David Wootton (Oxford: Clarendon Press, 1992).

107. On Descartes's debt to Charron, see especially José R. Maia Neto, *Academic Skepticism in Seventeenth-Century French Philosophy: The Charronian Legacy 1601–1662* (Heidelberg: Springer, 2014), 97–125.

108. Baillet, *Des-Cartes*, 1:86.

109. I follow the translation in Cole, *Olympian Dreams*, 32–40.

110. For further details on many of the events that follow, as well as an excel-

lent contextualization of German intellectual debates during Descartes's period in Germany, see Édouard Mehl, *Descartes en Allemagne 1619–1620: Le contexte Allemand de l'élaboration de la science Cartésienne* (Strasbourg: Presses Universitaires de Strasbourg, 2001).

111. Rodis-Lewis, *Descartes: Life and Thought*, 36.

112. Buzon, "Exemplaire."

113. Quoted in Cottingham, Stoothoff, and Murdoch, *Philosophical Writings*, 1:125.

114. Mehl (*Descartes en Allemagne*, 189) thinks it is probable.

115. Raz Chen-Morris, *Measuring Shadows: Kepler's Optics of Invisibility* (University Park: Pennsylvania State University Press, 2016).

116. Schuster, *Descartes Agonistes*, 153–63.

117. Ulinka Rublack, *The Astronomer and the Witch: Johannes Kepler's Fight for His Mother* (Oxford: Oxford University Press, 2015).

118. For a recent account, see Robert S. Westman, *The Copernican Question: Prognostication, Skepticism, and Celestial Order* (Berkeley: University of California Press, 2011), 492–94.

119. Pietro Redondi, *Galileo Heretic* (Princeton, NJ: Princeton University Press, 1987).

120. Mehl (*Descartes en Allemagne*, 190) thinks that Kepler was a kind of absent father for Descartes in sorting out the *Komentenstreit* of 1619.

121. Baillet, *Des-Cartes*, 1:68–69; translation of the section on Faulhaber in Watson, *Cogito*, 104–6; but Rodis-Lewis (*Descartes: Life and Thought*, 53) says that Rothen had died in 1617.

122. For a recent review of the details, see Joly, *Descartes et la chimie*, 52–58.

123. William R. Shea, "Descartes and the Rosicrucians," *Annali Dell'Instituto e Museo di Storia della Scienza di Firenze* 4 (1979): 34. Åkerman says that he published the book at Ulm and dedicated it to "Polybius Cosmopolita": Åkerman, *Rose Cross*, 221–22.

124. C. Adam and Tannery, *Oeuvres De Descartes*, 10:214; for an example of the argument that is was a satire, see Rodis-Lewis, *Descartes: Life and Thought*, 35.

125. Balzac to Descartes, March 30, 1628; on interpretations of the dreams, see Leroy, *Descartes: Philosophe au Masque*, 79–88, and Freud's interpretation at 89–90; Cole, *Olympian Dreams*; Alan Gabbey, "The Melon and the Dictionary: Reflections on Descartes's Dreams," *Journal of the History of Ideas* 59 (1998): 651–68.

126. The *Algemeine und General Reformation der gantzen weiten Welt* (*Universal and General Reformation of the Whole Wide World*) and the *Fama Fraternitatis* (*Echoes of the Fraternity*); see also Didier Kahn, "The Rosicrucian Hoax in France (1623–24)," in *Secrets of Nature: Astrology and Alchemy in Early Modern Europe*, ed. William R. Newman and Anthony Grafton (Cambridge, MA: MIT Press, 2001), 238.

127. Montgomery, *Cross and Crucible*, 165–68.

128. Govert Snoek, *De Rosenkruisers in Nederland: Voornamelijk in de eerste helft*

van de 17^e eeuw, een inventarisatie (Haarlem, Netherlands: Rozekruis Pers, 2006), 19–56.

129. The best recent case for the political importance of the Rosicrucian manifestos has been made by Åkerman, who in her *Rose Cross* has revised the views of Frances A. Yates, *The Rosicrucian Enlightenment* (London: Routledge and Kegan Paul, 1972).

130. This is the main argument of Montgomery, *Cross and Crucible*, vol. 1; for a careful general review of the Rosicrucian movement, see Donald R. Dickson, *The Tessera of Antila: Utopian Brotherhoods and Secret Societies in the Early Seventeenth Century* (Leiden: Brill, 1998), 18–88.

131. On chymical theory, see Jole Shackelford, *A Philosophical Path for Paracelsian Medicine: The Ideas, Intellectual Context, and Influence of Petrus Severinus (1540/2–1602)* (Copenhagen: Museum Tusculanum Press, 2004); Bruce T. Moran, *Distilling Knowledge: Alchemy, Chemistry, and the Scientific Revolution* (Cambridge, MA: Harvard University Press, 2005); William R. Newman, *Atoms and Alchemy: Chymistry and the Experimental Origins of the Scientific Revolution* (Chicago: University of Chicago Press, 2006); Lawrence M. Principe, ed., *Chymists and Chymisry: Studies in the History of Alchemy and Early Modern Chemistry* (Sagamore Beach, MA: Chemical Heritage Foundation and Science History Publications, 2007). On Paracelsus, see Walter Pagel, *Paracelsus: An Introduction to Philosophical Medicine in the Era of the Renaissance* (New York: S. Karger, 1958); Charles Webster, *Paracelsus: Medicine, Magic and Mission at the End of Time* (New Haven, CT: Yale University Press, 2008). On Germany, see Mehl, *Descartes en Allemagne*; Tara E. Nummedal, *Alchemy and Authority in the Holy Roman Empire* (Chicago: University of Chicago Press, 2007); Alisha Rankin, *Panaceia's Daughters: Noblewomen as Healers in Early Modern Germany* (Chicago: University of Chicago Press, 2013).

132. Wouter J. Hanegraaff, "Under the Mantle of Love: The Mystical Eroticisms of Marsillio Ficino and Giordano Bruno," in *Hidden Intercourse: Eros and Sexuality in the History of Western Esotericism*, ed. Jeffrey John Kripal and Wouter J. Hanegraaff (Leiden: Brill, 2008), 175–207.

133. Montgomery, *Cross and Crucible*, 39–66; Dickson, *Tessera of Antila*, 22–30.

134. Snoek, *De Rosenkruisers in Nederland*, 57.

135. Gérard Simon, *Kepler: Astronome Astrologue* (Paris: Gallimard, 1979); Michael Heyd, *"Be Sober and Reasonable": The Critique of Enthusiasm in the Seventeenth and Early Eighteenth Centuries* (Leiden: E. J. Brill, 1995), 112.

136. Marjorie Reeves, *Joachim of Fiore and the Prophetic Future* (New York: Harper and Row, 1977). As Didier Kahn notes, the beginning of the seventeenth century represents a high-water mark for interest in Lull: Kahn, "Rosicrucian Hoax," 251. Also see Tomás y Artau Carreras and Joaquín y Artau Carreras, *Historia de la Filosofía Española: Filosopfía Cristiana de los Siglos XIII al XV*, 2 vols. (Madrid: Real Academia de Ciencias Exactas, Fisicas y Naturales, 1939–43), 118–57.

137. C. Adam, *Vie et Oeuvres*, 29, note D.

138. Adam (ibid., 48) notes that Baillet translated the phrase as “Rien du tout” (nothing at all) when it should be “rien de certain” (nothing certain), which leaves the door open to Descartes’s own view of the brethren. On the “Studium Bonae Mentis,” see C. Adam and Tannery, *Oeuvres De Descartes*, 10:191–204.

139. Mehl (*Descartes en Allemagne*, 193) thinks that a reference in Faulhaber’s 1622 *Miracula Arithmetica* to his good friend, the noble (Herr) “Carolus Zolindius (Polybius),” is likely to be a reference to Descartes.

140. Shea, “Descartes and Rosicrucian Enlightenment.”

141. Borel, *Renatus Descartes*, 67.

Part Four

1. Quoted in Cottingham, Stoothoff, and Murdoch, *Philosophical Writings*, 1:115.

2. Tapie, *France in Age of Louis XIII*, 112.

3. “. . . dont quelques uns pourroient etre de sa connoissance”: Baillet, *Des-Cartes*,1:64.

4. Monnier, *Philippe de Béthune*, 49.

5. Baillet, *Des-Cartes*, 1:66.

6. Peter H. Wilson, *The Thirty Years War* (Cambridge, MA: Belknap Press of Harvard University Press, 2009), 299.

7. Baillet, *Des-Cartes*, 1:66–67.

8. Ibid., 70.

9. Martines, *Furies*.

10. P. Wilson, *Thirty Years War*, 299–308; Tapie, *France in Age of Louis XIII*, 113.

11. Germaine Lebel, *La France et les principautés Danubiennes (du XVIe siècle à la chute de Napoléon 1er)* (Paris: Presses Universitaires de France, 1955), 33–36.

12. Baillet, *Des-Cartes*, 1:73–75, 91.

13. Ibid., 1:70–73; Borel, *Renatus Descartes*, 8–9.

14. Baillet, *Des-Cartes*, 1:95; P. Wilson, *Thirty Years War*, 323.

15. Quoted in Cottingham, Stoothoff, and Murdoch, *Philosophical Writings*, 1:82.

16. P. Wilson, *Thirty Years War*, 324.

17. Quoted in Cottingham, Stoothoff, and Murdoch, *Philosophical Writings*, 1:403–4.

18. In *Discours*, quoted in ibid., 1:145.

19. Baillet, *Des-Cartes*, 1:95–97; quotation at 97.

20. The Poles had sent forces to help defeat the Bohemians, and the Turks had in turn entered in support of Bethlen; in 1621 the Ottoman Porte sent a powerful army against the Polish-Lithuanian Commonwealth that hoped to take the Ukraine and even reach the Baltic. But between the beginning of September and the beginning of October—in a conflict referred to as the battle of Khotyn or Chochim or Hotin—the commonwealth held off the Turks, leading to a treaty.

21. Baillet, *Des-Cartes*, 1:98–99.

22. E. Ladewig Petersen, in *The Thirty Years' War*, by Geoffrey Parker (London: Routledge and Kegan Paul, 1984), 72–73.

23. A branch of the family then ruling Pomerania descended from Anna of Lorraine, whose close relatives included the Prince of Orange and the Bourbons and Gonzagas.

24. Baillet, *Des-Cartes*, 102–3.

25. Ibid.

26. Wilson, *Thirty Years War*, 332.

27. C. Adam, *Vie et Oeuvres*, 62.

28. Edgar de Lanouvelle, *Gabrielle d'Estrées et les Bourbon-Vendôme* (Paris: Calmann-Lévy, 1936), 103.

29. Sutcliffe, *Guez de Balzac*, 23–27. The delegation included Richelieu and father Bérulle for the queen and the Cardinal de La Rochefoucauld and Philippe de Béthune for the king: Monnier, *Philippe de Béthune*, 47.

30. Sutcliffe, *Guez de Balzac*, 27.

31. Taipe, *France in Age of Louis XIII*, 124–25.

32. Baillet (*Des-Cartes*, 1:106) describes his properties as a fief named "le perron"—from which he took his title—a great house in Poitiers, and a market next to the house ("le marchais outré une maison"), as well as several farms ("arpens de terre labourable") near Availle (which is just outside of Châtellerrault).

33. Ibid., 1:106.

34. One antiquarian notes that "In 1623 the epidemic fevers in Europe became more fatal, as the period of pestilence [the summer] approached. Riverius, who has written on the epidemic fevers of this period in the south of France, observes that the mortality was great. . . . He refers to the city of Montpelier, where almost half died who were seized": Noah Webster, *A Brief History of Epidemic and Pestilential Diseases*, 2 vols. (1799; New York: Burt Franklin, 1970), 1:180. Didier Kahn ("Rosicrucian Hoax," 298–99) has found evidence that the plague was troublesome in Paris at the end of the summer. Colin Jones notes that the decade from 1625 to 1635 saw the largest number of plague tracts published in France, correlating with the worst period of the disease: Colin Jones, "Plague and Its Metaphors in Early Modern France," *Representations* 53 (1996): 103.

35. Baillet, *Des-Cartes*, 1:106: "l'on goûtoit le repos." On the date of the return of the court to Paris (January 10), see Kleinman, *Anne of Austria*, 60.

36. Although young (b. 1604), Claude Hardy was already accomplished and became a *conseiller* in the Châtelet de Paris; it is said that in 1623 Descartes resided in the house of Claude's father, Sébastien Hardy: Joseph-François Michaud et Louis-Gabriel Michaud, *Biographie universelle ancienne et moderne*, vol. 18 (Paris: Louis Vivés, 1857), 457.

37. Baillet, *Life of Descartes*, 54; the French is at Baillet, *Des-Cartes*, 1:118.

38. Baillet, *Life of Descartes*, 50; the French is at Baillet, *Des-Cartes*, 1:112.

39. Baillet, *Des-Cartes*, 115–16.

40. Schuster, *Descartes-Agonistes*, 164.

41. Didier Kahn, "Rosicrucian Hoax," 252.

42. Ibid., 264–68.

43. Baillet, *Des-Cartes*, 106–10.

44. Didier Kahn, "Rosicrucian Hoax," 244.

45. Howarth, *Life and Letters*, 18–19.

46. Jeffrey John Kripal and Wouter J. Hanegraaff, eds., *Hidden Intercourse: Eros and Sexuality in the History of Western Esotericism* (Leiden: Brill, 2008).

47. Montgomery, *Cross and Crucible*, 165. Boccalini's *Ragguagli di Parnaso* (*News-sheet from Parnassus*, 1612) mocks contemporary justice by having Apollo hear complaints about worldly decisions and hand down proper sentences instead. On the associations between philosophical libertinism and political reform in the Venice of the time, see Wilding, *Galileo's Idol*.

48. Didier Kahn, "Rosicrucian Hoax," 263, 268–75, 294.

49. *New Catholic Encyclopedia.*, s.v. "Duvergier de Hauranne," accessed May 15, 2017, http://www.newadvent.org/cathen/05218a.htm.

50. Didier Kahn, "Rosicrucian Hoax," 239–240; Snoek, *De Rosenkruisers in Nederland*, 19.

51. Didier Kahn, "Rosicrucian Hoax," 263, 268–75, 294.

52. Hayden, *France and Estates General*, 280.

53. John A. Lynn, *Giant of the Grand Siècle: The French Army, 1610–1715* (Cambridge: Cambridge University Press, 1997), 109.

54. Douglas Clark Baxter, *Servants of the Sword: French Intendants of the Army, 1630–70* (Urbana: University of Illinois Press, 1976), 4, 8, 12–14, 21, 78–79.

55. Baillet, *Des-Cartes*, 1:118.

56. P. Wilson, *Thirty Years War*, 381–84.

57. Baillet, *Des-Cartes*, 1:119.

58. See note 13 to part 1.

59. A. Cappelli, *Cronologia, Cronografia e Calendario Perpetuo* (Milan: Ulrico Hoepli, 1969).

60. Baillet, *Des-Cartes*, 120–21.

61. Anna Frank-van Westrienen, *De Groote Tour: Tekening van de Educatiereis der Nederlanders in de Zeventiende Eeuw* (Amsterdam: Noord-Hollandsche Uitgevers-maatschappij, 1983); Edward Chaney, *The Evolution of the Grand Tour: Anglo-Italian Cultural Relations since the Renaissance* (London: Frank Cass, 1998).

62. C. Adam, *Vie et Oeuvres*, 64; Gaukroger, *Descartes: Intellectual Biography*, 133; Rodis-Lewis, *Descartes: Life and Thought*, 59–61; Ariew et al., *Historical Dictionary of Descartes*, 4.

63. *Wikipedia*, s.v. "Liste des chevaliers de l'ordre du Saint-Esprit," last modified April 24, 2017, https://fr.wikipedia.org/wiki/Liste_des_chevaliers_de_l%27ordre_du_Saint-Esprit.

64. Monnier, *Philippe de Béthune*, 47–53.

65. Pierre, *Le Père Joseph*, 225.

66. Nancy Siraisi, Lucia Dacome, and Heikki Mikkeli have drawn our attention to Santorio in recent years, and there is currently an ongoing project to study his quantifying methods at the University of Exeter.

67. Ernst, *Campanella*, esp. 27, 220.

68. On the details of the printing and reception of *The Assayer*, see Redondi, *Galileo Heretic*, 36–51; Heilbron, *Galileo*, 263–65.

69. Stillman Drake, ed., *Discoveries and Opinions of Galileo* (Garden City, NY: Doubleday, 1957), 217–80. For a recent account of the book, see Heilbron, *Galileo*, 245–52.

70. Heilbron, *Galileo*, 266.

71. Francesco Barberini would become a Grand Inquisitor of the Roman Inquisition but did not vote against Galileo. On Guevara, see Annibale Fantoli, *The Case of Galileo: A Closed Question?* trans. George V. Coyne (Notre Dame, IN: University of Notre Dame Press, 2012), 141.

72. Borel, *Renatus Descartes*, 9.

73. See the letters to Mersenne, from the end of November 1633 to August 1634, in Cottingham, Stoothoff, and Murdoch, *Philosophical Writings*, 3:40–45.

74. Baillet, *Des-Cartes*, 1:124. Descartes's statement included: "Aussi ne vois-je rein dans ses livres qui me fasse envie, ni presque rien que je voulusse avoüer pour mien."

75. An even earlier version of the idea had been developed by Robert Grosseteste in the thirteenth century.

76. Baillet, *Des-Cartes*, c. 1:122: "M Descartes crut qu' il étoit bienséant à un gentihomme François d' aller render des civilitez à un cardinal neveu, destine pour faire dans on pays une fonction aussi importante qu' étoit cette legation."

77. Wilson, *Thirty Years War*, 383.

78. Baillet, *Des-Cartes*, 1:127.

79. Ibid., 129; C. Adam, *Vie et Oeuvres*, 69.

Part Five

1. For a rich and penetrating account of these developments, see Schuster, *Descartes-Agonistes*, 167–348.

2. On Villebressieu and Mersenne, see C. Adam, *Vie et Oeuvres*, 90–91; on Mersenne more generally, see Dear, *Mersenne*.

3. Baillet, *Des-Cartes*, 1:152.

4. Ibid., 1:132: "il étoit néantmoins trés-éloigné du libertinage."

5. Tapie, *France in Age of Louis XIII*, 172–73.

6. Gaukroger, *Descartes: Intellectual Biography*, 136.

7. Baillet, *Des-Cartes*, 1:140–41: Cardinal Barberini remained in Paris until September 1625.

8. Ibid., 1:134; Baillet, *Life*, 60–61.

9. Baillet, *Life*, 57.

10. Baillet, *Des-Cartes*, 1:136: the address is given as the rue Du Four Aux Trois Chappelets. Jouanna, *St. Bartholomew's Day Massacre*, 126.

11. Baillet, *Des-Cartes*, 1:153–54.

12. Harth, *Cartesian Women*, 26.

13. Constant, *Gaston d'Orleans*, 72–73.

14. For the case that Marie Le Jars de Gournay deserves the credit, see Ilsley, *Daughter of the Renaissance*, 230–31.

15. Harth, *Cartesian Women*, 26; C. Adam and Tannery, *Oeuvres De Descartes*, 12: 236. After the death of the duchesse du Montpensier, her uncle, Richelieu, intended Madame de Comballet to be married to Gaston d'Orléans: it never happened.

16. Constant, *Gaston d'Orleans*, 117.

17. Baillet, *Des-Cartes*, 1:144–45.

18. Ibid., 1:143: "Ce fut M Des Argues qui contribua principalement à le faire connoître au Cardinal de Richelieu: et quoi que M Descartes ne prétendît tirer aucun avantage de cette connoissance, il ne laissa pas de se reconnoître trés-obligé au zéle que M Des Argues faisoit paroître pour le servir."

19. Pierre Gatulle, *Gaston d'Orléans: Entre mécénat et impatience du pouvoir* (Seyssel, France: Champ Vallon, 2012), 98, 383, 377.

20. Baillet, *Des-Cartes*, 1:147; also see Pintard, *Libertinage Érudit*, 203–4.

21. Ilsley, *Daughter of the Renaissance*, 144.

22. Constant, *Gaston d'Orleans*, 15, 115, 202.

23. Sylvain Matton, "Cartésianisme et alchimie: à propos d'un témoignage ignoré sur les travaux alchimiques de Descartes. Avec une note sur Descartes et Gómez Pereira," in *Aspects de la tradition alchimique au XVIIe Siècl*[e] (Paris: Société d'Étude de l'Histoire de l'Alchimie, Archè, 1998), 53–54; also see Constant, *Gaston d'Orleans*, 202.

24. Frances Huemer, *Rubens and the Roman Circle: Studies of the First Decade* (New York: Garland, 1996), 87–91.

25. Father Bérulle would later be made cardinal.

26. For the biographical details of Bérulle's career, I follow Pierre, *Monarchie ecclésiale*, 346–49; on his work with the Carmelites, also see Jean Dagens, *Bérulle et les origines de la restauration Catholique (1575–1611)* (Paris: Desclée de Brouwer, 1952), 191–228.

27. Roger Ariew, "Oratorians and the Teaching of Cartesian Philosophy in Seventeenth-Century France," *History of Universities* 17 (2001–2): 47–80.

28. Gatulle, *Gaston d'Orléans*, 257–58.

29. For the report of his dressing in green taffeta, which comes from Le Vasseur, see C. Adam, *Vie et Oeuvres*, 74; on Anne of Austria's people, see Kleinman, *Anne of Austria*, 20; on Gaston as a "vert-galant," see Constant, *Gaston d'Orleans*, pp. 201–35.

30. Alexander Marr, "Crowned with Harmless Fire: A New Look at Descartes," *Times Literary Supplement*, March 13, 2015.

31. Jacques Thuillier, Barbara Brejon de Lavergnée, and Denis Lavalle, *Vouet: Galeries Nationales du Grand Palais, Paris, 6 Novembre 1990–11 Février 1991* (Paris: Editions de la Réunion des musées nationaux, 1990), 16.

32. Ibid., 100–108 for his major period in Rome (1622–27); Marr, "Crowned with Harmless Fire."

33. Gatulle, *Gaston d'Orléans*, 172.

34. "Il y aura plaisir à lire vos diverses aventures dans la moyenne et dans la plus haute region de l'air, à considerer vos prouesses contre les Geans de l'Escole, le chemin que vous avez tenu, *le progrez que vous avez fait dans le vérité* des choses, etc." Jean Louis Guez de Balzac to René Descartes, 30 March 1628, Circulation of Knowledge and Learned Practices in the 17th-Century Dutch Republic, accessed April 26, 2017, http://ckcc.huygens.knaw.nl/epistolarium/letter.html?id=desc004/1013.

35. For the association with Rosicrucianism, see G. Cohen, *Écrivains Français en Hollande*, 417, and Leroy, *Descartes: Philosophe au Masque*, 72–73; also see Didier Kahn, *Alchimie et Paracelsisme en France a la fin de la Renaissance (1567–1625)* (Geneva: Droz, 2007), 510 and Joly, *Descartes et la Chimie*, 61.

36. C. Adam and Tannery, *Oeuvres De Descartes*, 1:212–18; Cottingham, Stoothoff, and Murdoch, *Philosophical Writings*, 3:32–33.

37. On La Rochelle, see Borel, *Renatus Descartes*, 9; Baillet, *Des-Cartes*, 1:155–60; on a sample of doubts, see C. Adam, *Vie et Oeuvres*, 99; Rodis-Lewis, *Descartes: Life and Thought*, xii; Watson, *Cogito*, 140.

38. C. Adam, *Vie et Oeuvres*, 99.

39. Pierre Grillon, ed., *Les papiers de Richelieu: Section politique intérieure correspondence et papiers d'état*, 2 vols. (Paris: A. Pedone, 1977), 2:627, item 755: "Rolle des cap[itai]nes de l'equipage du Sr de la Richardière pour l'embarquement du secours de la citadelle de Saint-Martin de Ré." November 8, 1627: "Le capne Odart, admiral commandant le secours. Le capne Descart, visadmiral. . . ." Seventeen names are mentioned, with an eighteenth dismissed. A Protestant Sieur de la Richardiere can be found associated with Poitiers: "Histoires Singulaires," Overblog (blog), accessed April 26, 2017, http://histoiressingulieres.over-blog.com/pages/Quelques_filiations_protestantes_poitevines_3-6154880.html; and "Ingrand," Noms du Poitou (de la Pissarderie) (blog), June 22, 2013, http://nomsdupoitoudelapissarderie.blogspot.com/2013_06_01_archive.html.

40. Dutch naval support was required by the Treaty of Compiegne of June 1624, by which France subsidized the war against Spain in the low countries in return for naval support; the English were at that time allies of France, confirmed by the marriage treaty.

41. Gatulle, *Gaston d'Orléans*, 401.

42. Martines, *Furies*, 146–50.

43. I imagine that the note of November 8 is taken down at the end of the

action on the Isle de Ré. It is possible, however, that it simply indicates a resupply of troops following the conclusion of events on the island.

44. Baillet, *Des-Cartes*, 1:158–59.

45. Ibid., 1:157: "il se procura encore le plaisir de s' en entretenir avec les ingénieurs."

46. G. Cohen, *Écrivains Français en Hollande*, 430–32.

47. Baillet devotes an entire chapter to La Rochelle: Baillet, *Des-Cartes*, 2:155–60. Beeckman noted that after Descartes's October visit to him, Descartes promised to send him a copy of his geometry *from Paris*: Watson, *Cogito*, 141.

48. Ariew et al., *Historical Dictionary of Descartes*, quotations from p. 4.

49. Borel, *Renatus Descartes*, 9–11; Baillet, *Des-Cartes*, 2:160–66. On the documentation for the episode, which stems from Borel and Clerselier, see Watson, *Cogito*, 142–43.

50. C. Adam and Tannery, *Oeuvres De Descartes*, 1:212–18; Cottingham, Stoothoff, and Murdoch, *Philosophical Writings*, 3:32–33.

51. There is a technical issue with "right reason," which implies that all knowledge comes from the *logos*; like the Epicureans and others, Descartes considered the sources of knowledge of real things to lie in sense impressions rather than the *logos*. For an introduction to the issue, see Robert Hoopes, *Right Reason in the English Renaissance* (Cambridge, MA: Harvard University Press, 1962).

52. C. Adam and Tannery, *Oeuvres De Descartes*, 1:212–18; Cottingham, Stoothoff, and Murdoch, *Philosophical Writings*, 3:32–33.

53. While most commentators agree with Baillet, Henri Gouhier places the meeting between the autumn of 1627 and April or May 1628: Sylvain Matton, ed., *Lettres sur l'or potable, suives du traité de la connaissance des vrais principes de la nature et des mélanges, et de fragments d'un commentaire sur* L'amphitéâtre de la Sapience Éternalle *de Khunrath* (Paris: Société d'Étude de l'Histoire de l'Alchimie, Archè, 2012), 50.

54. Georg Lutz, *Kardinal Giovanni Francesco Guidi di Bagno: Politik und Religion im Zeitalter Richelieus und Urbans VIII* (Tübingen: Max Niemeyer, 1971), 547.

55. Matton, "Cartésianisme et Alchimie," 45; Monnier, *Philippe de Béthune*, 53.

56. Matton, *Lettres sur l'or potable*, 54.

57. Dedications of Jean Collesson (*L'idee parfaicte de la philosophie hermetique*, 1631) and Pierre Jean Fabre (*L'abrégé des secrets chymiques*, 1636); see Gatulle, *Gaston d'Orléons*, 257.

58. Jack A. Clarke, *Gabriel Naudé, 1600–1653* (Hamden, CT: Archon Books, 1970), 42; see also Cavaillé, *Dis/simulations*.

59. Gideon Manning, "Descartes and the Bologna Affair," *British Journal for the History of Science* 46, (2013): 1–13.

60. Baillet, *Des-Cartes*, 1:254; he also wrote of Descartes (1:253): "Il espéroit même que si opinions étoient jamais reçûës, toutes les controversies qui s' agitent dans la théologie pourroient tomber d' ells-mêmes, parce qu' ells sont fondées pour

la plûpart sur des principes de philosophie qu' il estimoit faux" ("He even hoped that if his opinions were ever accepted, all the controversies which agitated theology might fall of themselves, because they were founded for the most part on principles of philosophy which he considered false").

61. Constant, *Gaston d'Orleans*, 15, 115.

62. Chandoux was not a significant enough figure to draw any notice from Didier Kahn, *Alchimie*.

63. Commenting on Baillet's sources, see C. Adam, *Vie et Oeuvres*, 95.

64. Matton, "Cartésianisme et alchimie"; my great thanks to Dan Garber for alerting me to this reference.

65. Ibid., 16–21.

66. Ibid., 50–54.

67. Tapie, *France in Age of Louis XIII*, 172.

68. Rio Howard, "Guy de la Brosse: Botanique et chemie au début de la Révolution Scientifique," *Revue Histoire de Science* 31 (1978): 301–26; Rio Howard, "Medical Politics and the Founding of the Jardin Des Plantes in Paris," *Journal of the Society for the Bibliography of Natural History* 9 (1980): 395–402; Rio Howard, "Guy de la Brosse and the Jardin des Plantes in Paris," in *The Analytic Spirit*, ed. Harry Woolf (Ithaca, NY: Cornell University Press, 1981), 195–224.

69. Howard Solomon, *Public Welfare, Science and Propaganda in Seventeenth Century France: The Innovations of Théophraste Renaudot* (Princeton, NJ: Princeton University Press, 1972), esp. 14, 19, 20, 39–40; for the conferences, see Kathleen Wellman, *Making Science Social: The Conferences of Théophraste Renaudot 1633–1642* (Norman: University of Oklahoma Press, 2003).

70. Matton, "Cartésianisme et alchimie," 57–135.

71. Joly, *Descartes et la Chimie*.

72. Schuster, *Descartes-Agonistes*, quotation at 301, Schuster's analysis at 307–48, esp. 346–47.

73. Quotations in C. Adam and Tannery, *Oeuvres De Descartes*, 4:62, 78; Cottingham, Stoothoff, and Murdoch, *Philosophical Writings*, 1:143, 151; René Descartes to Marin Mersenne, 27 February 1637, in Cottingham, Stoothoff, and Murdoch, *Philosophical Writings*, 3:53. More generally, see Theo Verbeek, "Les passions et la fièvre: L'idée de la maladie chez Descartes et quelques Cartésiens Néerlandais," *Tractrix* 1 (1989): 45–61; Steven Shapin, "Descartes the Doctor: Rationalism and its Therapies," *British Journal for the History of Science* 33 (2000): 131–54.

74. Baillet, *Des-Cartes*, 1:165: "pria M Descartes qu' il pût l' entendre encore une autre fois sur le même sujet en particulier."

75. Baillet, *Des-Cartes*, 1:165–66.

76. C. Adam, *Vie et Oeuvres*, 96.

77. The suggestion is at ibid., 94.

78. Baillet, *Des-Cartes*, 1:165–66: "L' impression . . . se trouvent jointe à ce que son naturel et sa raison luy dictoient depuis long têms acheva de le déterminer. Jus-

ques là il n' avoit encore embrassé aucun parti dans la philosophie, et n' avoit point pris de secte, comme nous l' apprenons de luy meme."

79. Ibid., 1:166: "Il se confirma dans la resolution de conserver sa liberté, et de travailler sur la nature même sans s'arrêter à voir en quoi il s' approcheroit ou s' éloigneroit de ceux qui avoient traitté la philosophie avant lui."

80. Ibid.: "la chaleur du climat et la foule du grand monde."

81. Quotations from ibid.; on Picot, 168.

82. Balzac to Descartes, 30 March 1628 (see note 34 above).

83. One alternative, stressing the climate, is well presented by C. Adam (*Vie et Oeuvres*, 98): "Il était prêt même à écrire sa physique, et n'aura besoin pour cela que de deux ou trois années. Seulement il prot la résolution de quitter Paris et la France: il lui fallait la tranquillité d'une retraite à l campagne, sous un climat favorable. Il pensa ne trouver ce qu'il désirait, qu'à l'étranger, dans un pays qu'il connaissait déjà: l'Italie ne lui convenant pas, à cause du climat, il choisit de préférence la Holland" ("He was even ready to write his physics, and would need only two or three years. But he had come to a resolution of leaving Paris and France: he needed the tranquility of a retreat in the country, in a favorable climate. He thought he could only find what he desired abroad, in a country he already knew: Italy was not suitable to him because of the climate, so he chose Holland"). Another alternative, stressing the need for solitude, is Rodis-Lewis, Descartes: *Life and Thought*, 49: "at the end of 1628, he escaped once more into a solitude so perfect that his location has remained forever unknown."

84. Sawyer, *Printed Poison*, 141–42.

85. Quoted in Cottingham, Stoothoff, and Murdoch, *Philosophical Writings*, 1:396: "Passions," under the heading "self-satisfaction."

86. Tapié, *France in Age of Louis XIII*, 203.

87. Baillet, *Des-Cartes*, 1:177–78.

88. Ibid., 193–94.

89. Thomas M. Carr, "Descartes and Guez De Balzac: Humanist Eloquence Spurned," in *Descartes and the Resilience of Rhetoric: Varieties of Cartesian Rhetorical Theory* (Carbondale: Southern Illinois University Press, 1990). Also see Gournay's quarrel with Balzac over his *Lettres*, in Michèle Fogel, *Marie de Gournay: Itinéraires d'une femme savante* (Paris: Fayard, 2004), 238–40.

90. Howarth, *Life and Letters*, 19, 29.

91. Sutcliffe, *Guez de Balzac*, 30–31.

92. G. Cohen, *Écrivains Français en Hollande*, 417.

93. René Descartes, "Clarissimo Viro Domino: Censura quarundum Epistolarum Domini Balzacii," Circulation of Knowledge and Learned Practices in the 17th-Century Dutch Republic, accessed April 27, 2017, http://ckcc.huygens.knaw.nl/epistolarium/letter.html?id=desc004/1012.

94. Most studies of Descartes either ignore the episode or take it as merely a lit-

erary dispute, as in Carr, “Descartes and Guez De Balzac,” or Gaukroger, *Descartes: Intellectual Biography*, 181–82. Balzac’s biographers take it more seriously.

95. For this period of Balzac’s life, see Sutcliffe, *Guez de Balzac*, 30–32.

96. Lanouvelle, *Gabrielle d’Estrées*, 102–10. César was connected to the house of Lorraine by marriage to Françoise de Lorraine. After fleeing to Holland, he went on to England and did not return to France until the death of Richelieu; in 1643, he took part in the plot of the *cabale des importants*, one of whose leaders was Marie de Rohan.

97. Rodis-Lewis, *Descartes: Life and Thought*, 77, and 241n12: he made a move to recover the trunk only in 1634.

98. René Descartes to Marin Mersenne, 15 April 1630, in Cottingham, Stoothoff, and Murdoch, *Philosophical Writings*, 3:21, 23.

99. Opening of part 6, in C. Adam and Tannery, *Oeuvres De Descartes*, 6:60; Cottingham, Stoothoff, and Murdoch, *Philosophical Writings*, 1:141. We should imagine that he wrote these words while finishing his book, the contract for which was signed in 1636, for he elsewhere mentions the condemnation of Galileo as having occurred three years earlier (it took place in June 1633). The year 1628 is also consistent with his comment about how he had had his initial ambition nine years earlier still, i.e., 1619, which is when surviving notes about his workbook record that he had his famous set of three dreams leading to his resolution to search for the foundation of true knowledge.

100. René Descartes to Guez de Balzac, 15 April 1631, in Cottingham, Stoothoff, and Murdoch, *Philosophical Writings*, 3:30.

101. End of part 3 of the *Discours*, in C. Adam and Tannery, *Oeuvres De Descartes*, 6:30–31; Cottingham, Stoothoff, and Murdoch, *Philosophical Writings*, 1:126.

102. Descartes to Balzac, 15 April 1631.

103. Guez de Balzac, *Oeuvres*, ed. L. Moreau, 2 vols. (Paris: Jacques Lecoffre et Cie, 1854), 1:81; J. H. Elliott, *Richelieu and Olivares* (Cambridge: Cambridge University Press, 1984), 159.

104. Descartes to Balzac, 15 April 1631.

105. Ibid.

106. Ibid.

107. Sutcliffe, *Guez de Balzac*, 34–39.

108. “Jean-Louis Guez de Balzac: Biographie,” Académie française, accessed April 27, 2017, http://www.academie-francaise.fr/les-immortels/jean-louis-guez-de-balzac?fauteuil=28&election=13-03-1634.

109. René Descartes to Guez de Balzac, 5 May 1631, in Cottingham, Stoothoff, and Murdoch, *Philosophical Writings*, 3:31–32.

110. Ibid.

Part Six

1. Quoted in Cottingham, Stoothoff, and Murdoch, *Philosophical Writings*, 1:143.

2. Baillet, *Des-Cartes*, 2:67, 326; Ariew et al., *Dictionary*, 51.

3. Ferdinand Sassen, "De Reis van Pierre Gassendi in de Nederlanden (1628–1629)," in *Mededelingen der Koninklijke Nederlandse Akademie van Wetenschappen* (Amsterdam: N. V. Noord-Hollandsche Uitgevers Maatschappij, 1960), 263–307; Ferdinand Sassen, "De Reis van Marin Mersenne in de Nederlanden (1630)," in *Mededelingen van de Koninklijke Vlaamse Academie voor Wetenschappen Letteren en Schone Kunsten van België, Klasse der Letteren* 16, no. 4 (1964).

4. Descartes to Mersenne, 15 April 1630, in Cottingham, Stoothoff, and Murdoch, *Philosophical Writings*, 3:21.

5. Manning, "Descartes and Bologna Affair," 10; the business was reported aloud to the Bolognese senate on March 1633, with Torelli's initial effort occurring in late September 1632.

6. Tapié, *France in Age of Louis XIII*, 306–7.

7. The excitement of the moment is beautifully invoked in Redondi, *Galileo Heretic*.

8. C. Adam, *Vie et Oeuvres*, 587.

9. Bredius, *Johannes Torrentius Schilder*, 14–65; Snoek, *De Rosenkruisers in Nederland*, 105–68; see also Mike Dash, *Batavia's Graveyard: The True Story of the Mad Heretic Who Led History's Bloodiest Mutiny* (London: Weidenfeld and Nicolson, 2002), 35–43, 239–51.

10. On Beeckman's student, see Berkel, *Isaac Beeckman*, 111; on the association of Van Hogelande and Torrentius with Thibault's academy, see Dash, *Batavia's Graveyard*, 34.

11. Bredius, *Johannes Torrentius Schilder*, 9.

12. C. Wilson, "Descartes and Corporeal Mind.".

13. Quoted in Cottingham, Stoothoff, and Murdoch, *Philosophical Writings*, 3:317; Cottingham, *Descartes' Conversation*, 47, 33.

14. Snoek, *De Rosenkruisers in Nederland*, 153. See Verbeek, *La Querelle d'Utrecht*; Theo Verbeek, *Descartes and the Dutch: Early Reactions to Cartesian Philosophy, 1637–1650* (Carbondale: Southern Illinois University Press, 1992).

15. Verbeek, *Descartes and Dutch*, 57.

16. G. Cohen, *Écrivains Français en Hollande*, 636.

17. Ibid., 636–67; from Baillet, *Des-Cartes*, 2:327: "de ses grandes mérites et de l'utilité que sa Philosophie et les recherches de ses longues etudes procuroient au genre humain; comme aussi pour l'aider à continuer ses belles experiences qui requeroient de la dépense."

18. For his complaints about the bad treatment he received concerning this "parchment," see René Descartes to Pierre Chanut, 31 March 1649, in Cottingham, Stoothoff, and Murdoch, *Philosophical Writings*, 3:371.

19. G. Cohen, *Écrivains Français en Hollande*, 641.

20. Ibid., 645; Frederick Henry had died on March 14, 1647.

21. Batiffol, *Duchesse du Chevreuse*, 323–33.

22. G. Cohen, *Écrivains Français en Hollande*, 646.

23. Shorto, *Descartes' Bones*.

24. He signed the baptismal register in the Reformed church in Deventer under the name of "Reyner Jochems" (René, son of Joachim): Rodis-Lewis, *Descartes: Life and Thought*, 245; C. Adam and Tannery, *Oeuvres de Descartes*, 1:367, April 27, 1637.

25. On his later reputation, see, for example, Stéphane Van Damme, *Descartes* (Paris: Presses de Sciences Politiques, 2002); Antonio Negri, *The Political Descartes: Reason, Ideology and the Bourgeois Project*, trans. Matteo Mandarini and Alberto Toscano (London: Verso, 2007).

26. Mullin, "If Truth Were Like Money."

27. Andrade, *Gunpowder Age*.

INDEX

Page numbers in italics refer to figures.